T0207518

Kölner Beiträge zur Didaktik der Mathematik

In dieser Reihe werden ausgewählte, hervorragende Forschungsarbeiten zum Lernen und Lehren von Mathematik publiziert. Thematisch wird sich eine breite Spanne von rekonstruktiver Grundlagenforschung bis zu konstruktiver Entwicklungsforschung ergeben. Gemeinsames Anliegen der Arbeiten ist ein tiefgreifendes Verständnis insbesondere mathematischer Lehr- und Lernprozesse, auch um diese weiterentwickeln zu können. Die Mitglieder des Institutes sind in diversen Bereichen der Erforschung und Vermittlung mathematischen Wissens tätig und sorgen entsprechend für einen weiten Gegenstandsbereich: von vorschulischen Erfahrungen bis zu Weiterbildungen nach dem Studium.
Diese Reihe ist die Fortführung der „Kölner Beiträge zur Didaktik der Mathematik und der Naturwissenschaften“.

Lukas Baumanns

Mathematical Problem Posing

Conceptual Considerations and Empirical Investigations for Understanding the Process of Problem Posing

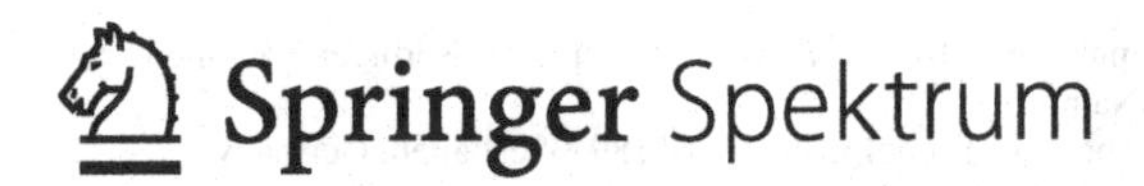

Lukas Baumanns
Cologne, Germany

Die vorliegende Veröffentlichung wurde von der Mathematisch-Naturwissenschaftlichen Fakultät der Universität zu Köln als Dissertation angenommen. Am 26. April 2022 fand die Disputation am Institut für Mathematikdidaktik statt. Die Prüfungskommission bestand aus den Gutachtern Prof. Dr. Benjamin Rott und Prof. Dr. Hans-Georg Weigand sowie dem Prüfungsvorsitzenden Prof. Dr. André Bresges und dem Beisitzer PD Dr. Stephan Berendonk.

ISSN 2661-8257 ISSN 2661-8265 (electronic)
Kölner Beiträge zur Didaktik der Mathematik
ISBN 978-3-658-39916-0 ISBN 978-3-658-39917-7 (eBook)
https://doi.org/10.1007/978-3-658-39917-7

Responsible Editor: Marija Kojic
This Springer Spektrum imprint is published by the registered company Springer Fachmedien Wiesbaden GmbH, part of Springer Nature.
The registered company address is: Abraham-Lincoln-Str. 46, 65189 Wiesbaden, Germany

Foreword

What a ride.

Who could have known what a ride it was going to be.

Back in 2017, when I talked with this bright young man about his Master thesis, it seemed like an easy task: Replicate the method I used in my PhD thesis in the field of problem solving and transfer it to problem posing to characterize problem-posing processes. The result was one of the best Master theses I had supervised, but we quickly realized that our work was not finished, but had just started. Fortunately, that bright young man continued to do research on mathematical problem posing by starting a PhD project. But instead of immediately returning to the analyses of problem-posing processes, the terminology had to be addressed as this field of research is still developing and in need of proper definitions. The results were not one, but two review articles. Review articles are hard to publish; in the first tries, they both got rejected, and rejected again, discouraging our endeavors. But eventually, they both got published in very good journals. Then, the empirical work continued with innovative methods and results that finally led to another two articles—with ups and downs in the revision processes—being published in very prestigious journals. These four articles, together with others texts, form the basis of the PhD thesis in front of the readers. Watch out for the author of this thesis—I am convinced that Lukas' ideas will continue to shake up research on mathematical problem posing.

What a ride, indeed.

What a joyride!

Cologne in August of 2022,
Prof. Dr. Benjamin Rott

Danksagung

„Da ligget se!“

Voller Stolz präsentiere ich den aktuellen Stand meiner Forschung zum Prozess des Aufwerfens mathematischer Probleme. Diesen Stand und die Freude an der Arbeit an diesem Projekt habe ich unter anderem den folgenden, großartigen Menschen zu verdanken.

Zunächst möchte ich Prof. Dr. Benjamin Rott danken. Es war über die gesamten viereinhalb Jahre stets ein Privileg, Teil deiner Arbeitsgruppe zu sein, die Möglichkeit zu haben, mit dir zu diskutieren und Gedankengänge zu präzisieren. Vielen Dank dafür, Benjamin!

Weiterer Dank gilt Prof. Dr. Hans-Georg Weigand, dem ich für sein flexibles Engagement im Vorfeld der Disputation für die Übernahme des Zweitgutachtens danken möchte.

Außerdem danke ich natürlich allen aktuellen und ehemaligen Mitgliedern unserer Arbeitsgruppe für unzählige Diskussionen und Denkanstöße: Julia, Raja, Anne, Udil, Moritz, Janine, Anna, Peter und Simon. So sind die entscheidenen Gedanken des zweiten Artikels, zum Beispiel, in einer verzweifelten, aber enorm hilfreichen AG-Sitzung entstanden. Ich danke euch für die gemeinsame Zeit sehr und vermisse die Treffen mit euch unheimlich.

Des Weiteren danke ich dem gesamten Institut für Mathematikdidaktik an der Universität zu Köln für die letzten vier Jahre. Die Diskussionen innerhalb der Mitarbeiter*innenseminare war immer ein hilfreicher Blick aus teils ganz anderen Blasen, die der eigenen Betriebsblindheit gut entgegenwirkten. Auch die gemeinsamen Mittagessen vor allem mit

Stephan und Stefan waren immer eine angenehme Möglichkeit, Gedanken auszutauschen, kreisen oder schweifen zu lassen.

Auch den zahlreichen Personen innerhalb der Mathematikdidaktik, die ich in den vergangenen Jahren im Rahmen meiner Tätigkeit kennenlernen durfte und von denen ich für diese Arbeit viel gelernt und darüber hinaus viel Freundschaft erfahren habe, möchte ich herzlich danken. Auch wenn ich hier auf eine namentliche Aufzählung verzichte, wisst ihr hoffentlich, dass ihr gemeint seid. Besonderer Dank gilt jedoch in diesem Zusammenhang einer Person, deren Wert ich in den letzten Jahren beruflich, privat und weit darüber hinaus gar nicht hoch genug schätzen kann, mein ständiger Begleiter. Danke, Max!

Ich danke auch allen meinen Freund*innen, deretwegen ich Gott sei Dank länger mit dem Fertigstellen meiner Dissertation gebraucht habe. Ohne euch wäre mein Leben nur halb so lustig und aufregend. Ohne alle Namen aufzuzählen, wisst auch ihr sicherlich, dass ihr gemeint seid. Ich danke selbstverständlich *nicht* der COVID19-Pandemie, dessentwegen ich diese Freund*innen viele Monate nicht sehen konnte und keine andere Wahl hatte, als Teile der durch die Freund*innen glücklicherweise verlorenen Zeit wieder aufzuholen.

Ich möchte auch allen Forschenden rund um das Thema Problem Posing danken, deren Gedanken und Untersuchungen stets Quelle neuer Erkenntnisse für mich waren (insbesondere für die Artikel 1 und 2). Außerdem danke ich allen Studierenden, deren Problem-Posing-Prozesse Anstoß und Ursprung meiner empirischen Studien waren (insbesondere für die Artikel 3 und 4).

Abschließend danke ich meinen Eltern und meinen Brüdern, bei denen ich mich in manchen Phasen der Promotion zu ihrer Besorgnis zu selten gemeldet habe, für eure ständige und bedingungslose Unterstützung. Ohne euch wäre ich nicht(s). Danke, Mama und Papa. Und danke, Markus und Niklas. Ich liebe euch.

Abstract

Mathematical problem posing as the substantive formulation of mathematical problems is an activity that lies at the heart of mathematics. In recent years, research in mathematics education has endeavored to gain insights into problem posing—conceptually as well as empirically. In problem-posing research, there has been a focus on analyzing products, that is, the posed problems. Insights into the processes that lead to these products, however, have so far been lacking. Within four journal articles, summarized in this cumulative dissertation, the author attempts to contribute to the understanding of problem-posing processes through conceptual considerations and empirical investigations. The conceptual part consists of a conducted systematic literature review to investigate problem-posing situations and problem-posing activities. The studies in the empirical part deal with the analyses of problem-posing processes of pre-service mathematics teachers from a macroscopic and microscopic perspective. The aim is to develop coherent and meaningful conceptual perspectives for analyzing empirical observations of problem-posing processes.

Contents

Part I

Introduction

1 Motivation

> "The mathematical experience of the student is incomplete if he never had an opportunity to solve a *problem invented by himself.*" (Pólya, 1957, p. 68)

In his seminal work on mathematical problem solving, Pólya emphasizes the importance of the experience of solving one's own problems. In mathematics education research, the activity of finding and formulating a problem is called *problem posing.*

This thesis focuses on the process of problem posing. The aim is to contribute to the understanding of this process through conceptual considerations and empirical investigations. This introduction, therefore, (1) argues for the necessity of *conceptual work* in the research field of problem posing as well as (2) motivates the focus on *problem-posing processes* in our empirical analyses.

Early conceptual work on problem posing can be found especially in the 1980s and 1990s. Their focus was on the development of purposeful problem-posing strategies (Brown & Walter, 1983; Kilpatrick, 1987), the description of different activities and cognitive processes that are referred to as problem posing (Silver, 1994), the relationship to related constructs such as mathematical creativity and problem solving (Silver, 1994; Silver, 1997), or the categorization of problem-posing situations (Stoyanova & Ellerton, 1996). These seminal conceptual considerations were followed by a cascade of empirical studies based on them. These studies investigated students' and prospective teachers' procedures and

L. Baumanns, *Mathematical Problem Posing*, Kölner Beiträge zur
Didaktik der Mathematik, https://doi.org/10.1007/978-3-658-39917-7_1

difficulties in posing problem (English, 1998; Silver et al., 1996), cross-national differences in the relationship between problem-posing and problem-solving abilities (Cai & Hwang, 2002), or the relationship between problem posing and mathematical creativity (Yuan & Sriraman, 2011; Leung, 1997).

With this plethora of empirical studies, it seems that the diversity of research questions pursued and empirical results obtained within these studies outweighs the existing theoretical and conceptual frameworks available for the analysis of the respective data. About 20 years after his seminal article on problem posing in 1994, Silver (2013) reflects:

> "At this time, I think our compilation of empirical evidence regarding mathematical problem posing has far outstripped our available theoretical tools to analyze and interpret the evidence." (p. 160)

The present thesis is, among other things, concerned with the development of such conceptual tools for analyses in the field of problem posing. The focus is in particular on the analysis of problem-posing processes. We want to concretize what we mean by problem-posing processes by looking at processes and products in mathematics education in general and in the field of problem posing in particular.

In mathematics education research in general, there is a long history of analyzing products and/or processes. Products are static states and results of a (learning) process. With regard to problem posing, the products are understood to be problems that have arisen as a result of a problem-posing process. The mathematical problem-posing process is defined as a purposeful activity of posing and formulating a problem in the field of mathematics. While products are often more accessible for analysis, the construction of new knowledge emerges in the process. As Freudenthal (1991) states, "the use of and the emphasis on *processes* is a *didactical principle.* Indeed, didactics itself is concerned with processes" (p. 87, emphases in original).

In problem-posing research, we see a lot of methodical work on analyzing problem-posing products (Bicer et al., 2020; Bonotto, 2013; Leung, 1993a; Singer et al., 2017; Van Harpen & Presmeg, 2013; Van Harpen & Sriraman, 2013; Voica & Singer, 2013). Within these studies, problem-posing products are mostly analyzed to determine participants' problem-posing abilities. There are many established frameworks to analyze the products of problem-posing processes: Based on the proposal of Silver (1997), some studies analyze the number, the diversity of ideas used, and the originality in terms of rarity of posed problems (Bonotto, 2013; Van Harpen & Presmeg, 2013). This approach originates from creativity research (Guilford, 1967; Torrance, 1974) and is also used to capture mathematical creativity in the context of multiple solution tasks in problem solving (Leikin & Lev, 2013). Another type of product analysis can be found in Silver and Cai (1996). They categorize posed problems first in terms of whether it is a mathematical problem or not. Afterwards they categorize the mathematical problems in terms of whether or not the mathematical question is ultimately solvable. Other studies also use this framework for their analyses (Bonotto & Santo, 2015; Kwek, 2015; Xie & Masingila, 2017). Bonotto and Santo (2015) adapted this framework by analyzing whether the posed problems are plausible or have sufficient information to solve it.

Despite the importance of mathematical learning processes suggested above, only a small number of studies investigate problem-posing processes (Kontorovich & Koichu, 2016; Kontorovich et al., 2012; Kontorovich & Koichu, 2012; Leung, 1994; Pelczer & Gamboa, 2009; Patáková, 2014). In these studies, individual problem-posing processes are often described and analyzed. For example, Kontorovich and Koichu (2016) had the opportunity to observe an expert problem poser for mathematics competitions such as the International Mathematical Olympiad pose problems. Although they are reticent to draw conclusions due to the limited scope of the study, they identify certain mechanisms of problem-posing experts. For example, the expert problem poser draws new problems from a family of familiar problems. To date, however, currently little research exists on the general and fundamental processes that constitute the activity of problem posing (Cai et al., 2015). This dissertation attempts to contribute to this field of research on problem-posing processes. The detailed objectives of the conceptual considerations and empirical investigations of this dissertation is outlined in section 2 on the facing page.

Broken down, the overall objective of this PhD project is to better understand problem-posing processes. Thus, the general orientation of this PhD project in mathematics education research is to better understand an inherently mathematical activity. To situate this work generally, we refer to Schoenfeld (2000). According to Schoenfeld (2000), mathematics education research has – similar to mathematics research – two purposes: (1) *Pure* mathematics education research aims at understanding the nature of mathematical thinking, teaching, and learning. (2) *Applied* mathematics education research aims at using these understandings to improve mathematics instruction. The present work is located in the tradition of purpose (1), pure mathematics education research. We try to better understand the activity of problem posing by means of content-related conceptual considerations as well as empirical-based process observations. With this aim to analyze and better understand mathematical activities as well as typical mathematical thinking processes involved in problem posing, the present work is located in the core of mathematics education research (Wittmann, 1995).

2 Objectives

As problem posing still is a field which is in need of definition and conceptual understanding (Silver, 2013; Singer et al., 2013), the main goal of this contribution is to develop coherent and meaningful conceptual frameworks that help organize empirical observations on problem-posing processes (Shapira, 2011). The four publications contributing to mathematics education research as part of this PhD project can be divided into two parts: (1) a conceptual part consisting of two journal articles and (2) an empirical part consisting of two journal articles. Their goals for the general aim (i.e., contribute to the understanding of the problem-posing process through conceptual considerations and empirical investigations) are summarized in the following:

(1) Conceptual part:
Before dealing empirically with problem-posing processes, a systematic literature review was conducted to investigate problem-posing *situations* and problem-posing *activities.*

Journal article 1:
Before we look at problem-posing activities reported in studies in high-ranked journals on mathematics education, we will look at the initiators of such processes, namely problem-posing situations. Article 1 is a review of 271 systematically collected problem-posing situations. This review aims to uncover differences regarding the openness of various problem-posing situations. As problem-posing situations initiate problem-posing activities, this review was a necessary preliminary step before

L. Baumanns, *Mathematical Problem Posing*, Kölner Beiträge zur
Didaktik der Mathematik, https://doi.org/10.1007/978-3-658-39917-7_2

taking a more detailed look at problem-posing activities in the further course of this thesis.

Journal article 2:
Article 2 is a systematic review of problem-posing activities reported in empirical studies in high-ranked journals on mathematics education. We examine problem-posing activities initiated by the open problem-posing situations conceptualized in Article 1. Article 2 aims to characterize different problem-posing activities based on three different dimensions.

(2) Empirical part:
Within two articles, the empirical part deals with the analysis of 32 problem-posing processes of pre-service mathematics teachers.

Journal article 3:
Article 3 reports the results of an empirical study of 32 problem-posing processes of pre-service mathematics teachers who were observed during problem posing. The aim of the study is to develop a phase model for the macroscopic description of problem-posing processes. We want to identify recurring and distinguishable activities of pre-service teachers dealing with problem-posing situations.

Journal article 4:
Within the macroscopic analyses of the study reported in Article 3, microscopic expressions have been observed that provide an important contribution to the understanding of problem-posing processes. These are expressions concerning the regulation of the process at a metacognitive level. The aim of Article 4 is therefore to characterize problem-posing-specific aspects of metacognitive behavior in problem-posing processes.

Table 2.1 on the next page summarizes the journal articles that are part of this publication-based dissertation. This table concisely summarizes the contribution of each article.

In addition to these core articles of this publication-based dissertation, three other publications of this work are noteworthy as part of this project. Their contribution as well as their purpose in the context of the present PhD project are outlined in the following:

- In a book chapter, problem posing is illustrated from a theory-based subject-matter didactics perspective with four selected mathematical examples. The different mathematical natures of problem posing are illuminated in mathematical terms, which could not be addressed in the four core journal articles. Findings from this chapter can be found in section 4.2 of the theoretical frame.
- A fifth journal article summarizes definitions of problem posing, its relations to other constructs, and its implementation into research and teaching settings. Thus, we extend the conceptual approaches of the core journal articles of this PhD project with a theoretical synthesis of problem-posing definitions, relations to other constructs, and its methodical implementation.
- Within a practical article, findings from our conceptual and empirical research are applied in the context of a teaching sequence on the generalization of the Pythagorean theorem. The aim is to show how the conceptual and empirical findings of the present PhD project can be useful for implementation in school practice.

Table 2.2 on page 11 summarizes the contributions of these additional publications related to this PhD project.

	Journal article 1	**Journal article 2**	**Journal article 3**	**Journal article 4**
Authors	Baumanns, L. & Rott, B.	Baumanns, L. & Rott, B.	Baumanns, L. & Rott, B.	Baumanns, L. & Rott, B.
Title	Rethinking problem-posing situations: A review	Developing a framework for characterizing problem-posing activities: A review	The process of problem posing: Development of a descriptive phase model of problem posing	Identifying metacognitive behavior in problem-posing activities. Development of a framework and a proof of concept
Journal	*Investigation in Mathematics Learning*	*Research in Mathematics Education*	*Educational Studies in Mathematics*	*International Journal of Science and Mathematics Education*
Contribution	Systematic literature review on the openness of problem-posing *situations* used in empirical studies in high-ranked journals on mathematics education	Systematic review of problem-posing *activities* reported in empirical studies in high ranked journals on mathematics education using three dimensions	Development a phase model for describing problem-posing processes by (1) identifying recurring and distinguishable activities in problem-posing processes and (2) deriving a general structure (i.e., sequence of distinguishable activities) of the observed processes	Investigating metacognitive behavior in problem posing by (1) identifying problem-posing specific aspects of metacognitive behaviors and (2) applying these identified metacognitive behaviors to illustrate differences in problem-posing processes

Tab. 2.1.: Overview of the four articles within this publication-based dissertation[a]

[a] We would like to mention at this point that Articles 1 and 2 are written in British English and Articles 3 and 4 are written in American English due to the specifications of the respective publishers.

	Book chapter	**Journal article 5**	**Practice article**
Author/s	Baumanns, L.	Papadopoulos, I., Patsiala, N., Baumanns, L. & Rott, B.	Schreck, A. & Baumanns, L.
Title	Four mathematical miniatures on problem posing	Multiple approaches to problem posing: Theoretical considerations to its definition, conceptualization, and implementation	Nicht verzagen, selbst was fragen! Digitale Hilfsmittel als adaptives Werkzeug
Journal/Book	*Problem Posing and Solving for Mathematically Gifted and Interested Students – Best Practices, Research and Enrichment*, Springer	*CEPS Journal*	*mathematik lehren*
Contribution	Illustrating mathematical problem posing from a subject-matter didactics perspective using four mathematical miniatures as an example	Capture different meanings and aspects of problem posing by approaching it from three different levels: (1) by comparing definitions, (2) by relating it to other constructs, and (3) by referring to research and teaching settings	Illustrate the path from the Pythagorean theorem to the cosine theorem using problem posing supported by dynamic geometry software for classroom use

Tab. 2.2.: Overview of the additional articles and book chapters referred to in this publication-based dissertation

3 Thesis Structure

The structure of the present work is illustrated in Figure 3.1 on page 15.

The main section begins with a theoretical frame. This theoretical frame aims to situate the activity of problem posing in mathematical and mathematics-educational terms. For this purpose, the theoretical frame begins with a subject-matter didactics analysis of the mathematical activity of problem posing. Problem posing is first illustrated with a geometric example. Subsequently, problem posing is conceptually specified based on definitions and conceptions within mathematics education. On this basis, four different problem-posing activities are then illustrated, each using a mathematical example. In this context, empirical findings of these four mathematical problem-posing activities are also summarized.

After this subject-matter didactics analysis, problem posing is subsequently located more generally within mathematics education. For this purpose, the constructs of modelling, proving, problem solving, as well as creativity and giftedness, which are related to problem posing, are considered. All these constructs are first defined and conceptualized, then their connection to problem posing is highlighted, and finally empirical findings on these connections are summarized.

The main body of the thesis contains four core journal articles. These articles are, as described above, divided into a conceptual and an empirical part. The conceptual part includes the systematic review of the problem-posing situations as well as the systematic review of problem-posing activities. The empirical part includes the macroscopic development

L. Baumanns, *Mathematical Problem Posing*, Kölner Beiträge zur
Didaktik der Mathematik, https://doi.org/10.1007/978-3-658-39917-7_3

of a phase model and the microscopic identification of metacognitive behavior in problem-posing processes.

In a concluding outlook, the insights of the four journal articles are summarized. Finally, interests of future research subsequent to this project are summarized based on findings across all four journal articles.

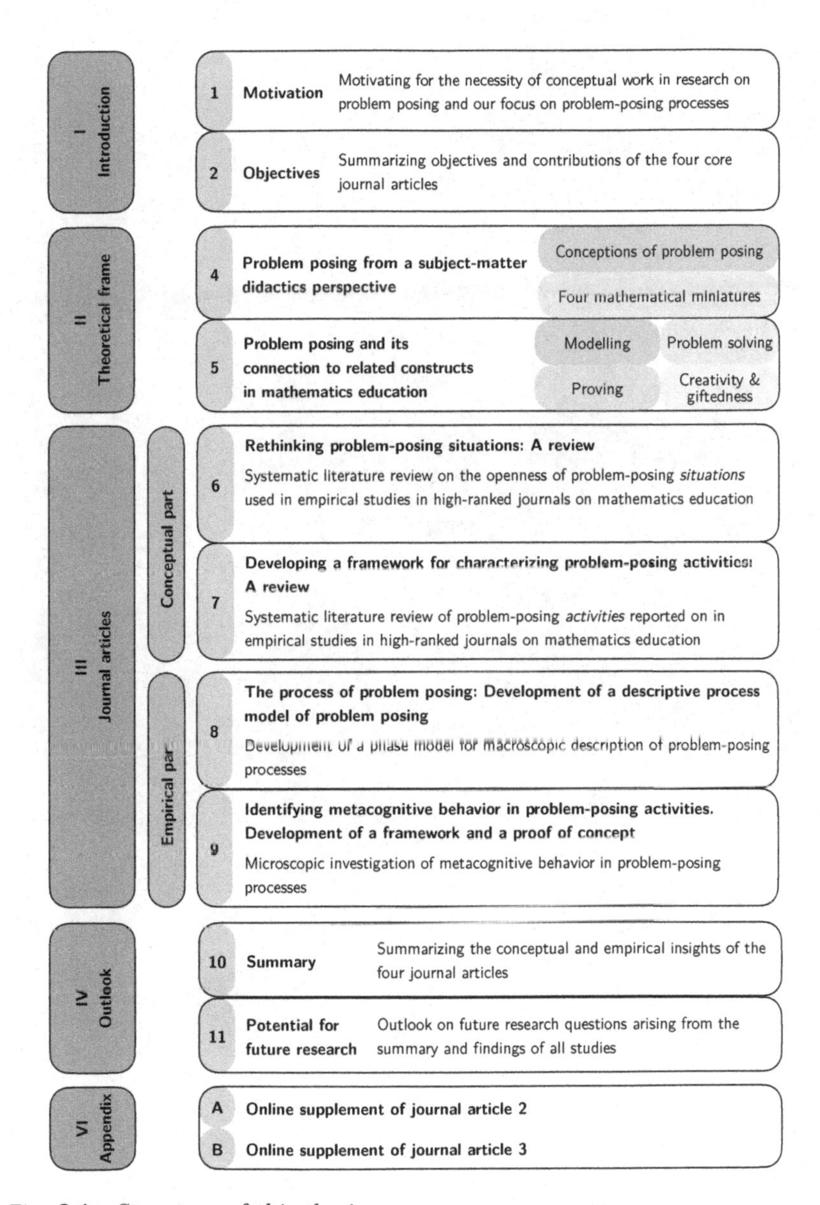

Fig. 3.1.: Structure of this thesis

Part II

Theoretical frame

Advanced organizer

In the following sections we will approach the activity of problem posing. To explain the structure of this section, let us use a metaphor: Imagine problem posing is a house in a housing estate. The purpose of this chapter is to learn about this house, its construction, foundation and rooms, observe residents in it and scout the neighborhood. In other words: We want to understand problem posing as a mathematical activity by looking at the different faces of it, what empirical insights there is on it, and what related constructs from mathematics education can help in conceptualizing problem posing in the context of this thesis.

First, we will look at the house as a whole from the outside. So, we will first look at problem posing on the basis of a specific example (see chapter 4.1 on page 21). Then we want to look at the fundament on which the house is built. So, we will deal with different definitions and conceptualizations of problem posing within mathematics education (see chapter 4.2 on page 35). Now that we have examined the house in its general form, we want to have a look inside and walk into the different rooms. This means that – based on the different definitions and conceptualizations – we will look at different types of problem posing. To do this, we look into each room separately, see what interior is in it and what load bearing walls were drawn on the fundament. So, in each case we look at the corresponding problem-posing type in a mathematical miniature, as well as findings from empirical studies on these problem-posing types (see chapter 4.2.1 to chapter 4.2.4).

Once the problem-posing house in detail, we will take a look at its neighborhood. We want to look at three different houses in this neighborhood, the *modelling* house, the *proving* house, the *problem-solving* house, and the *creativity & giftedness* house. We look at all three houses from the outside and describe the streets leading to our problem-posing house. Thus, we want to draw conceptual connections between other constructs of mathematics education that are close to problem posing in their research tradition (see chapter 5 on page 67).

During this house tour and exploration of the neighborhood, we also want to look at how people live in the problem-posing house, how they behave in the different rooms, and what their path to the neighboring houses looks like. In other words: We look more closely at trends in empirical research on problem posing and summarize their findings.

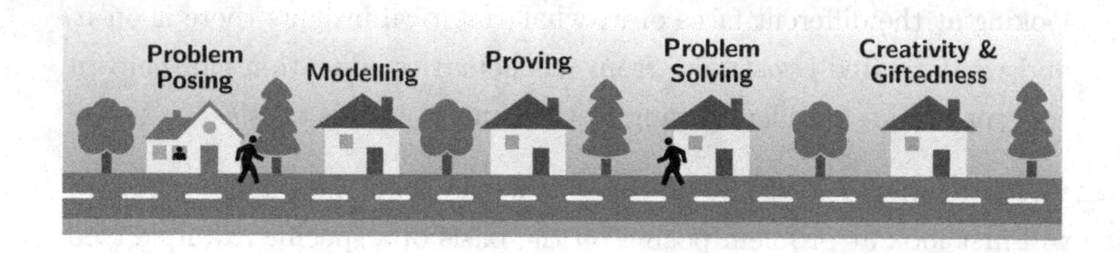

4 Problem posing from a subject-matter didactics perspective

4.1 Illustration with a geometric example

In this chapter, we will first consider problem posing using an illustrative example from geometry. The starting point of this illustration is a theorem from elementary geometry. Problem posing is used in this chapter with two goals: (1) to prove the presented theorem and (2) to generalize the presented theorem. Thereby we want to authentically illustrate a typical mathematician's activity, namely finding and varying problems, i.e. problem posing.

The way to the theorem is through the following problem:

> **Initial problem 1:**
> Construct squares on each side of a parallelogram $ABCD$. What figure is created when you connect the centers of these squares?

A quick picture should help to grasp this problem.

L. Baumanns, *Mathematical Problem Posing*, Kölner Beiträge zur Didaktik der Mathematik, https://doi.org/10.1007/978-3-658-39917-7_4

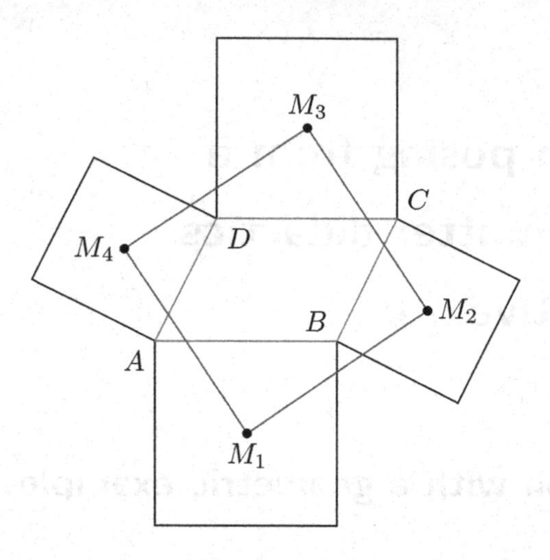

So, the centers M_1, M_2, M_3, M_4 of the constructed squares look like they are vertices of a square themselves! As mathematicians, however, the pure optics should not yet convince us. Let us formulate a theorem that is to be proved.

> **Theorem 1:**
> Construct squares on each side of a parallelogram $ABCD$. Their centers form a square.

$M_1M_2M_3M_4$ is a square if and only if all sides have the same length and all angles are the same size which is $90°$. To show the equality of these sizes, one approach would be to show the congruence of suitable triangles. For this, we draw auxiliary lines. Since we have already drawn the centers of the squares, let us use the center of the parallelogram P, which is the intersection point of the diagonals. We now connect P with M_1, M_2, M_3, and M_4 to get four triangles PM_1M_2, PM_2M_3, PM_3M_4, and PM_4M_1. Can we show the congruence of these triangles?

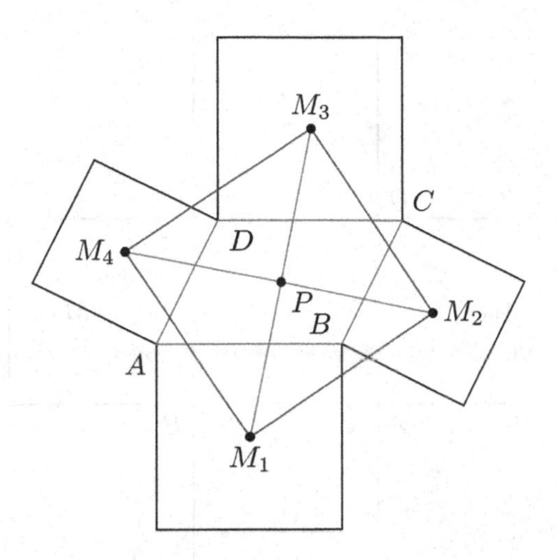

To prove the congruence of these triangles, we want to show that $|\overline{PM_1}| = |\overline{PM_3}|$. As P is the center of point symmetry of the parallelogram $ABCD$, this will probably be true, but I cannot come up with a better justification. And even if, the proof is still missing that also the interior angles are 90°. I am therefore following Pólya's adive to *pose a simpler theorem* where the solution is easier and hope that this will help me proof theorem 1.

If our conjecture is correct that the centers of the squares over the sides of a parallelogram are vertices of a square, then this should also be true for a very special parallelogram, namely the square:

> **Theorem 2:**
> Construct squares on each side of a square $ABCD$. Their centers form a square.

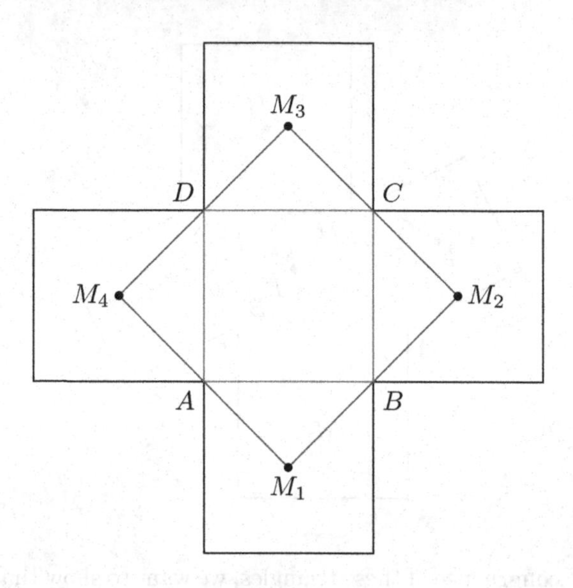

As expected, $M_1M_2M_3M_4$ looks like a square again. There are five congruent squares in this drawing, thus, the lines $\overline{M_1M_2}$, $\overline{M_2M_3}$, $\overline{M_3M_4}$, and $\overline{M_4M_1}$ go along the diagonals of the erected squares through the vertices of the square $ABCD$. This simplification provides the triangles AM_1B, BM_2C, CM_3D, and DM_4A. If we can now show that (1) $A \in \overline{M_4M_1}$, $B \in \overline{M_1M_2}$, $C \in \overline{M_2M_3}$, and $D \in \overline{M_3M_4}$, (2) the triangles AM_1B, BM_2C, CM_3D, and DM_4A are congruent, and (3) are right-angled at their respective vertices M_i for $i \in \{1, 2, 3, 4\}$ (then the rhombus is a square).

(1) All five visible squares are congruent and are lined up like a chessboard. Therefore, the connections of diagonally adjacent squares lie on their diagonals and thus run through their vertices.

(2) We show exemplarily the congruence of AM_1B and BM_2C. The remaining argumentations work analogously. $|\overline{AB}| = |\overline{BC}|$ by definition. As argued in (1), $\overline{M_1M_2}$ lies on the diagonals of the constructed squares of $\overline{AB}$ and $\overline{BC}$. This can analogously be

applied to $\overline{M_1M_4}$ and $\overline{M_2M_3}$. Thus, $|\sphericalangle BAM_1| = |\sphericalangle M_1BA| = |\sphericalangle M_2BC| = |\sphericalangle BCM_2| = 45°$. Thus, the triangles $A \in M_4M_1$ and $B \in M_1M_2$ coincide in one side as well as its adjacent angles. They are therefore congruent. So, $M_1M_2M_3M_4$ is a rhombus.

(3) Since $|\sphericalangle BAM_1| = |\sphericalangle M_1BA| = 45°$, $|\sphericalangle BM_1A| = 90°$. The same applies for the other triangles BM_2C, CM_3D, and DM_4A. The rhombus $M_1M_2M_3M_4$ is, therefore, a square.

Now that we have proven theorem 2, let us look at theorem 1 again. In the proof for theorem 2, the lines from the centers of constructed squares to the vertices of $ABCD$ helped us in finding suitable triangles to examine congruence. So we now draw these lines as auxiliary lines in our picture for theorem 1:

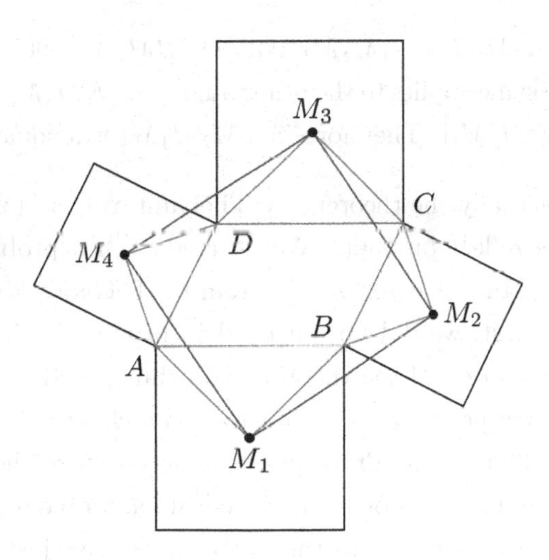

These new lines form four triangles, that is AM_4M_1, BM_1M_2, CM_2M_3, and DM_3M_4, which share one side with our alleged square. We now want to show that (1) these triangles are congruent (then $M_1M_2M_3M_4$ is a rhombus) and that (2) $|\sphericalangle M_2M_1M_4| = |\sphericalangle M_1M_4M_3| = |\sphericalangle M_4M_3M_2| = |\sphericalangle M_3M_2M_1| = 90°$ (then the rhombus is a square).

(1) We show exemplarily $AM_1M_4 \cong BM_1M_2$. The other congruences work analogously. Obviously, $|\overline{AM_1}| = |\overline{BM_1}|$ and $|\overline{AM_4}| = |\overline{BM_2}|$, because the lengths are half diagonals of identical or congruent squares. Now for the enclosed angles:

$$\begin{aligned}
|\sphericalangle M_1AM_4| &= |\sphericalangle M_1AB| + |\sphericalangle BAD| + |\sphericalangle DAM_4| \\
&= 45^\circ + 180^\circ - |\sphericalangle CBA| + 45^\circ = 270^\circ - |\sphericalangle CBA| \\
&= 360^\circ - 45^\circ - |\sphericalangle CBA| - 45^\circ \\
&= 360^\circ - (|\sphericalangle M_2BC| + |\sphericalangle CBA| + |\sphericalangle ABM_1|) \\
&= |\sphericalangle M_2BM_1|
\end{aligned}$$

Thus, AM_1M_4 and BM_1M_2 are congruent and, therefore, $M_1M_2M_3M_4$ is a rhombus.

(2) Since $|\sphericalangle M_2M_1B| = |\sphericalangle M_4M_1A|$ and $|\sphericalangle BM_1A| = 90^\circ$, also $|\sphericalangle M_2M_1M_4| = 90^\circ$. The same applies to the other angles $|\sphericalangle M_1M_4M_3|$, $|\sphericalangle M_4M_3M_2|$, and $|\sphericalangle M_3M_2M_1|$. Therefore, $M_1M_2M_3M_4$ is a square!

Theorem 1 is actually the theorem by Thèbault-Yaglom (Yaglom, 1962, p. 96). Let us reflect on that. We started with a problem that we wanted to solve. In this case, the theorem by Thèbault-Yaglom was to be proved. At first, we did not succeed in proving it. Therefore, we specialized a condition of the theorem and thus posed a new problem. This simplification provided us with a way to tackle the initial theorem again and actually prove it. Problem posing has occurred here as helpful *for* problem solving. We now want to consider problem posing *as an end in itself*, and go deeper into the problem we have just solved.

Now what if we didn't have a parallelogram as an initial figure, but a trapezoid?

Problem 2:
Construct squares on each side of a trapezoid $ABCD$. What figure is created when you connect the centers of these squares?

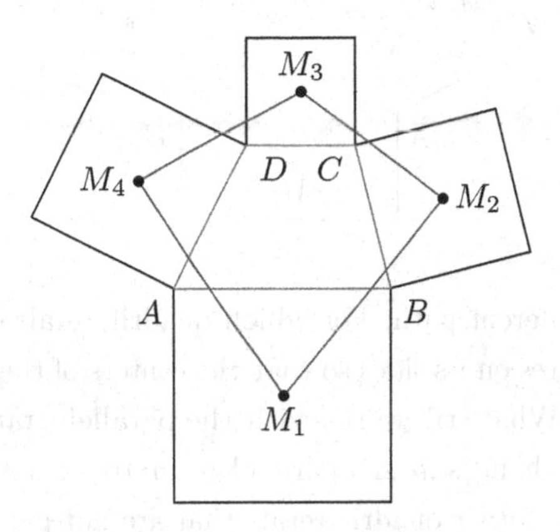

This actually does not look like a square, nor like another special quadrilateral. So, how can we now obtain the square? For this, we must probably change the constructed squares on the sides of the trapezoid cleverly. A solution to the question of which quadrilaterals on the sides of the trapezoid would have to be constructed is provided by Berendonk (2019):

Theorem 3:
„If you erect squares on the two ‚nonparallel' sides of a trapezoid and a rectangle on each of the two parallel sides, such that its ‚height' is equal to the length of the opposite side of the trapezoid, then the centers of the four erected quadrangles will form the vertices of a square." (Berendonk, 2019)

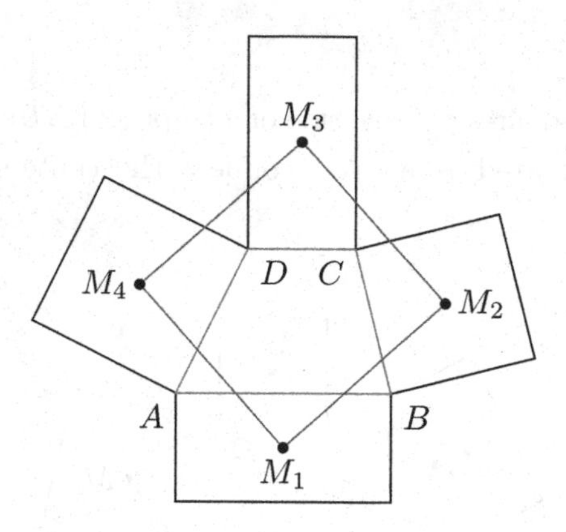

We will take a different path: For which quadrilaterals can we continue to construct squares on its sides so that the centers of these squares make a square again? What still works with the parallelogram fails with the trapezoid. What changes from a parallelogram to a trapezoid? If you apply this relation to other quadrilaterals that are not parallelograms, you find that it fails. Thus, the parallelogram is characterized by something specific, which seems to be relevant for the property investigated here. The parallelogram is the only point-symmetric quadrilateral. Maybe it is connected with this. Let us consider a simpler figure to get a better grasp of this relationship, namely the triangle:

> **Problem 3:**
> Construct equilateral triangles on each side of an equilateral triangle ABC. What figure is created when you connect the centers of these equilateral triangles?

More generally speaking, the *center* is the barycenter, so the center of gravity.

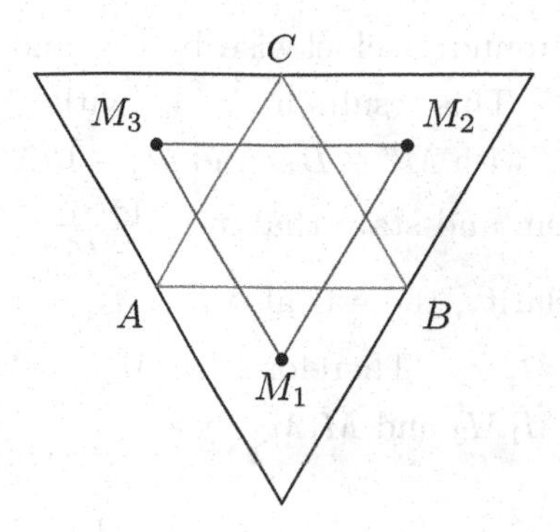

So, this obviously works and we will refrain from providing proof at this point. Now, what happens if we change triangle ABC by moving C slightly so that an arbitrary triangle is formed?

> **Theorem 3:**
> Construct equilateral triangles on each side of any triangle ABC. Their barycenters form an equilateral triangle.

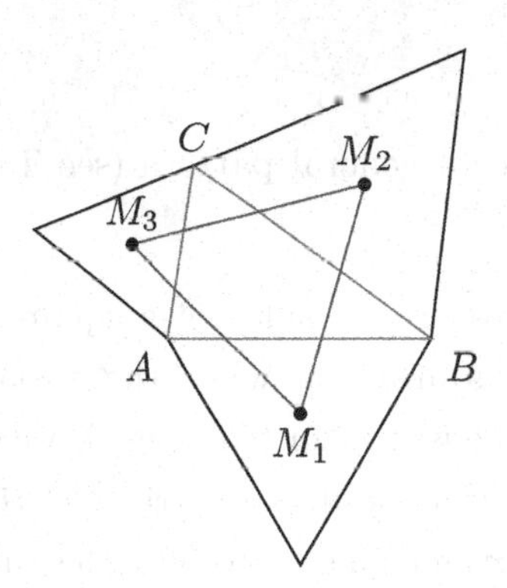

Amazingly, $M_1M_2M_3$ seems to remain an equilateral triangle! This result is called Napoleon's Theorem (first proven by Fischer (1863)). We will prove this theorem by showing that $|\overline{M_1M_3}| = |\overline{M_1M_2}|$ and $|\sphericalangle M_2M_1M_3| = 60°$. This proof follows Smyth (2007).

We rotate $\overline{M_1M_3}$ around A clockwise by $30°$ and $\overline{M_1M_3}$ around B anticlockwise by $30°$. This results in $\overline{M_1'M_3'}$ with $M_1' \in \overline{AD}$ and $M_3' \in \overline{AC}$ as well as $\overline{M_1''M_2'}$ with $M_1'' \in \overline{BD}$ and $M_2' \in \overline{BC}$. We can now apply the intercept theorem and state that $\tau := \frac{|\overline{AM_1'}|}{|\overline{AD}|} = \frac{|\overline{AM_3'}|}{|\overline{AC}|} = \frac{|\overline{BM_1''}|}{|\overline{BD}|} = \frac{|\overline{BM_2'}|}{|\overline{BC}|}$. Due to similarity, there is also $|\overline{M_1'M_3'}| = \tau \cdot |\overline{CD}| = |\overline{M_1''M_2'}|$ and $\overline{M_1'M_3'} \parallel \overline{CD} \parallel \overline{M_1''M_2'}$. Therefore, $|\overline{M_1M_3}| = |\overline{M_1M_3}|$. Due to the twofold rotation of $\overline{M_1M_3}$ and $\overline{M_1M_3}$ by $30°$, $|\sphericalangle M_2M_1M_3| = 60°$.

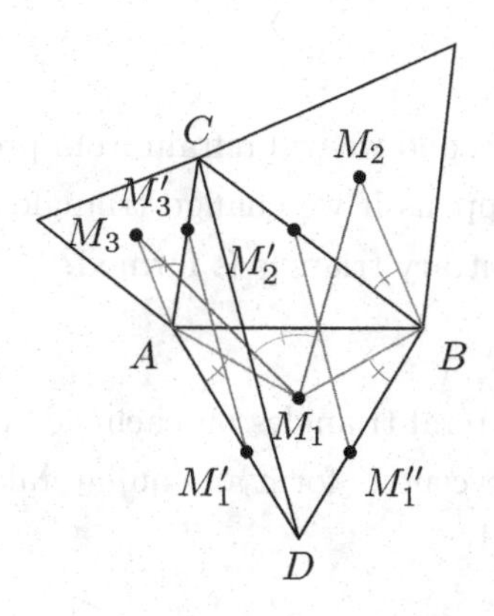

Now we are able to see a general pattern (see Table 4.1 on the next page).

If we construct squares on the sides of a square, their centers form a square. This remains invariant when the initial square becomes a parallelogram. The same is true for triangles. If we construct equilateral triangles on the sides of an equilateral triangle, their centers form an equilateral triangle. This remains invariant if the initial triangle becomes an arbitrary triangle. It remains to reason that this pattern will continue. We will at this point assume without proof that if you erect regular n-gons on each side of a regular n-gon, the barycenters form a regular n-gon. For the initial square, this property remained when the initial square becomes a parallelogram. It also remained when the initial equilateral

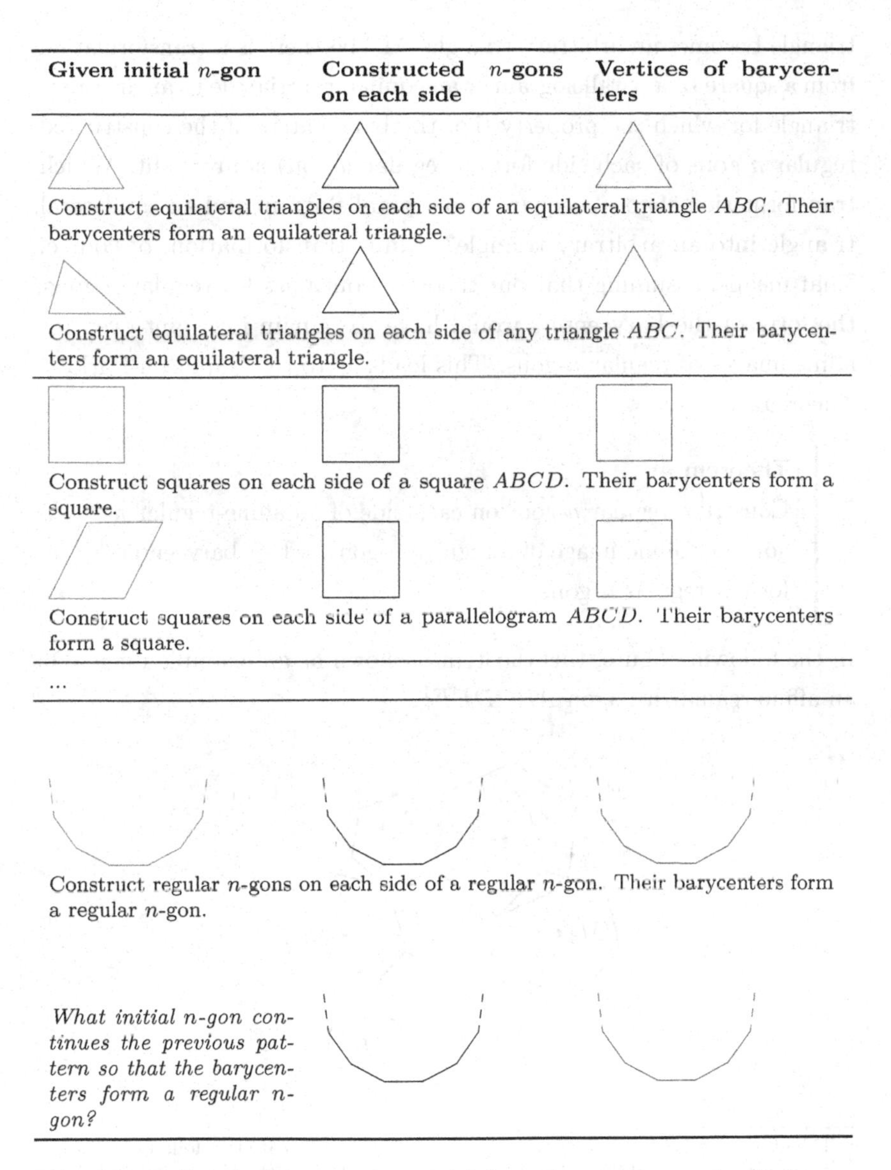

Given initial n-gon	**Constructed n-gons on each side**	**Vertices of barycenters**
Construct equilateral triangles on each side of an equilateral triangle ABC. Their barycenters form an equilateral triangle.		
Construct equilateral triangles on each side of any triangle ABC. Their barycenters form an equilateral triangle.		
Construct squares on each side of a square $ABCD$. Their barycenters form a square.		
Construct squares on each side of a parallelogram $ABCD$. Their barycenters form a square.		
...		
Construct regular n-gons on each side of a regular n-gon. Their barycenters form a regular n-gon.		
What initial n-gon continues the previous pattern so that the barycenters form a regular n-gon?		

Tab. 4.1.: Relations between given initial regular n-gon, constructed regular n-gons on each side of the initial regular n-gon, and form of barycenters of the constructed regular n-gons with regard to initial problem 1.

triangle becomes an arbitrary triangle. Maybe there is a transformation from a square to a parallelogram or an equilateral triangle to an arbitrary triangle for which the property (i.e. the barycenters of the constructed regular n-gons of each side form a regular n-gon) is invariant. Which transformation shapes a square into a parallelogram and an equilateral triangle into an arbitrary triangle? Affine transformation, of course! That means, assuming that our theorem continues for regular n-gons, the form of the barycenters would have to remain invariant even for affine images of regular n-gons. This leads us to the Napoleon-Barlotti Theorem:

Theorem 4:
Construct regular n-gons on each side of an affine-regular n-gon (i.e. affine image of a regular n-gon). Their barycenters form a regular n-gon.

In the following figure, this theorem is shown as an example for $n = 6$, an affine-regular hexagon $ABCDEF$[1].

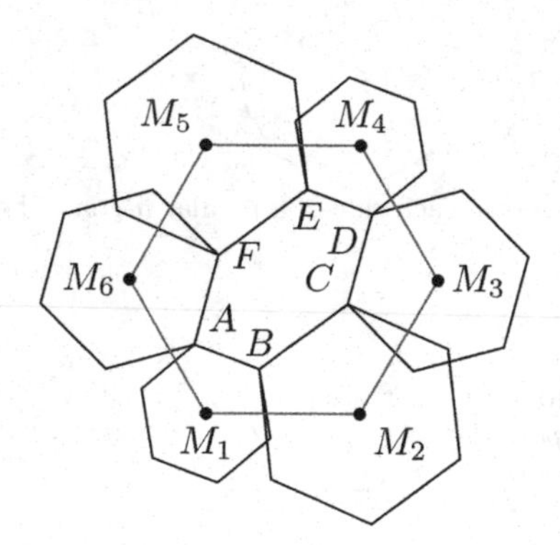

[1] The affine-regular hexagon $ABCDEF$ in the figure is created as follows: We take a regular hexagon $A'B'C'D'E'F'$ on the vertex $\overline{A'B'}$ with $A' = (0,0)$ and $B' = (1,0)$. The matrix underlying the affine transformation is $M = \begin{pmatrix} 0,8 & 0,8 \\ -0,3 & 1,1 \end{pmatrix}$. B, for example, is created through the following calculation: $B = A + M \cdot (B - A) = M \cdot B = \begin{pmatrix} 0,8 & 0,8 \\ -0,3 & 1,1 \end{pmatrix} \cdot \begin{pmatrix} 1 \\ 0 \end{pmatrix} = \begin{pmatrix} 0.8 \\ -0.3 \end{pmatrix}$. The same applies to A, C, D, E, and F.

The barycenters actually appear to be vertices of a regular hexagon! Due to the complexity of the proof, we will omit it here. We refer to the work of Fisher et al. (1981).

Let us reflect again on what has happened in this journey. After proving initial problem 1, we wanted to go deeper into it, so we started to vary individual conditions of the theorem. To do this, we first changed a parallelogram into a trapezoid. We found that the property that the barycenters form a square did not persist. To restore this, we changed another condition of the initial problem, namely the constructed squares on the sides of the respective quadrilateral. With cleverly selected rectangles, we were able to create a square here as well (Berendonk, 2019). However, we wanted to stay with the squares on the sides and see for which quadrilaterals the property that the barycenters form a square remains invariant. To understand this better, we specialized the theorem again by considering regular triangles instead of regular quadrilaterals. We found that when we construct equilateral triangles on each side of an equilateral triangle, their barycenters form an equilateral triangle. And even if the initial triangle is arbitrary, the barycenters of constructed equilateral triangles on each side are vertices of an equilateral triangle. By reflecting on the conditions we modified during problem posing, this surprising finding helped identify a general pattern. The function that transforms a regular triangle into an arbitrary triangle and a square into a parallelogram is the affine transformation (e.g., shear or squeeze transformation). Assuming that this finding proves true, we could generalize the initial problem for all affine-regular n-gons. These latter theorems remain unproven due to their complexity, but now offer the opportunity for further exploration.

We hint at other paths that could have been taken in connection with the modification of the initial theorem. In problem 2 we saw that for the trapezoid the barycenters of the constructed squares on each side do not form a square.[2] What if we now change the constructed figures on each side – something we did not change in our main path above? In general, we do not find a noteworthy insight for every trapezoid. However, if we consider very specific quadrilaterals, namely inscribed quadrilaterals, we get the following insight: Construct rectangles on each side of an inscribed quadrilateral so that one side of the rectangle coincides with one side of the inscribed quadrilateral and the other side of the rectangle is exactly as long as the respective opposite side in the inscribed quadrilateral. The barycenters of these rectangles form another rectangle (see Vargyas, 2020).

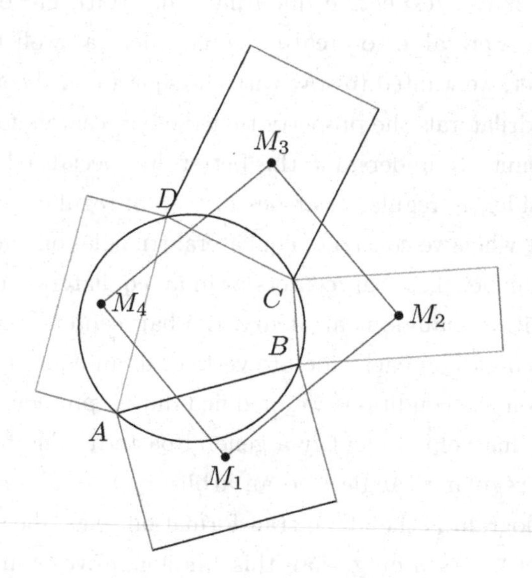

One can also generalize the dimension of the theorem. If one constructs cubes with congruent bases on each face of a cube, will the centers of the constructed cubes eventually form an octahedron? As we can see,

[2] In fact, the diagonals of the quadrilaterals in the case of the trapezoid in problem 2 intersect in a right angle. Further, the diagonals have the same length. Thus, the quadrilateral is orthodiagonal and equidiagonal. Surprisingly, this applies to *any* quadrilateral. This theorem is also known as van Aubel's Theorem (Nishiyama, 2011).

there are numerous possibilities to explore this theorem further. We want to leave it here as an introduction and reflect on the activity of problem posing.

The activity of problem posing, which is the focus of this work, had at least two different functions here. In the beginning, problem posing lead to a simpler theorem that was easier to tackle and helped solving the original problem. Afterwards, problem posing served to gain a more general insight into the relations of the geometric objects in the initial problem. These are only two of numerous functions that problem posing can fulfill. These functions do not only apply to the field of geometry. Problem posing can be pursued, as we will see in the coming chapters, in many areas of mathematics. In this section, we have practically conceptualized these functions. In the following sections, we want to define problem posing theoretically in a broader way.

4.2 Conceptions of problem posing – Four mathematical miniatures

This section attempts to summarize definitions and conceptions of problem posing and using these insights to illustrate problem-posing activities through mathematical miniatures from a subject-matter didactic perspective. This section has the following aims:

1. To identify problem-posing activities based on definitions and concepts from mathematics education.
2. To present these identified problem-posing activities through selected mathematical miniatures.

Larger sections of chapter 4.2 will be published in "Sarikaya, D., Heuer, K., Baumanns, L., & Rott, B. (Eds.) (2022). *Problem Posing and Solving for Mathematically Gifted and Interested Students – Best Practices, Research and Enrichment.* Springer."

3. To summarize the results of empirical research within the identified problem-posing activities.

Figure 4.1 summarizes the structure of this section: First, key definitions and conceptions of problem posing from mathematics education are summarized. From these definitions and conceptions, four different problem-posing activities are elaborated based on the different goals problem posers pursue: (1) *Problem posing as generating new problems*, (2) *problem posing as reformulating a given problem for problem solving*, (3) *problem posing as reformulating a given problem for investigation*, and (4) *problem posing as constructing tasks for others*. Each of these problem-posing activities is first illustrated with a mathematical miniature. Findings of empirical studies dealing with the respective problem-posing activity are then summarized.

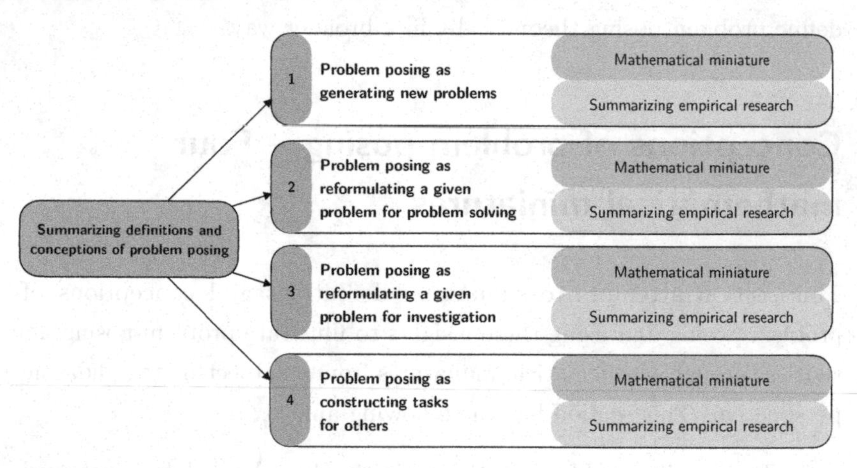

Fig. 4.1.: Structure of this section

Posing questions has always been a central part of constructing new knowledge for scientific progress but also for the individual learning process. Already Einstein and Infeld (1938) state:

> “Galileo formulated the problem of determining the velocity of light, but did not solve it. The formulation of a problem is often more essential than its solution, which may be merely a matter of mathematical or experimental skill. To raise new questions, new possibilities, to regard old problems from a new angle, requires creative imagination and marks real advance in science.” (p. 95)

This can also be applied to mathematics in which posing problems in itself represents a central value. Cantor (1867) already stated as the third thesis of his dissertation: “In re mathematica ars proponendi quaestionem pluris facienda est quam solvendi” (p. 26). This translates as: In mathematics, the art of posing a question is of greater value than solving it. We find similar statements from Lang (1989, p. 70), Poincaré (1973, p. 246), or Hilbert (1900, p. 262) who also emphasize the importance of posing problems in the field of mathematics. For example, Erdős is considered to be one of the most prolific mathematicians who, in addition to solving many problems, was also, and perhaps especially, known for posing problems in various fields. Indeed, it were his problems and not necessarily their solutions that opened up new areas of number theory, discrete mathematics, and graph theory (Ramanujam, 2013). In fact, he gave prize money for numerous problems, which was set higher depending on the estimated difficulty.

In mathematics education research, posing problems has been mentioned at least since dealing with problem solving. Problem posing in this context is seen as a partial aspect of problem solving. In his short dictionary of heuristics, Pólya (1957) describes at least two functions of problem posing. On the one hand, he describes problem posing for problem solving by varying the problem (pp. 209). Pólya gives several examples of how varying a problem can help to solve it. Specifically, as one example in this section of his seminal book *How to solve it*, he solved the problem of constructing a trapezoid using only the four given sides by posing two auxiliary problems through variation, making the trapezoid first a triangle and finally a parallelogram. We also saw this kind of problem posing in section 4.1 when we specialized initial problem 1 in order to prove it.

On the other hand, Pólya asks the question "Can you use the result, or the method, for some other problem?" (pp. 64). With this, he addresses that one should reap the fruits after solving a problem. In this context, he describes three paths for new problems in terms of their accessibility and interestingness. (1) Using the canonical variation strategies generalization, specialization, analogy, decomposing, and recombining, learners can get to new problems. However, these problems are rarely accessible as they soon get too difficult. We observed that in section 4.1 when especially the final generalization of the initial theorem was too difficult to access as we could not use a previous idea in order to prove the Napoleon-Barlotti Theorem. (2) It is also possible to pose new problems, which can be solved using the same method that was used to solve the initial problem. However, according to Pólya, these are rarely interesting as there are no structural new insights. (3) What is difficult is finding new problems that are accessible but remain interesting. Pólya provides a few examples where this endeavor has been successful.

Schoenfeld takes up Pólya's ideas in his seminal remarks on problem solving (1985b, 1989, 1992). As he states, problem posing helps for problem solving, identical to what we have already seen in Pólya (1957). Schoenfeld suggests, in the context of problem solving, that if difficulties arise, one should use modification to arrive at simpler problems (1985b, pp. 102).

In mathematics education research, the focused investigation of problem posing – and thus the terminological definition of this concept of activity – began in the 1980s (Walter & Brown, 1977; Butts, 1980; Brown & Walter, 1983; Ellerton, 1986; Kilpatrick, 1987). Butts (1980) pointed out that the way problems are posed significantly affects the problem solver's motivation to solve the problem, as well as his or her comprehension of the key concepts underlying the problem. Of particular note is certainly Brown and Walter's well-known "What-If-Not" strategy (Brown & Walter, 1983; Brown & Walter, 2005). This strategy involves going through five stages:

Level 0 Choosing a Starting Point

Level 1 Listing Attributes

Level 2 What-If-Not-ing

Level 3 Question Asking or Problem Posing

Level 4 Analyzing the Problem

The starting point can be a problem, a theorem, or even any kind of material (e.g., a geoboard, see section 4.2.1 on page 50) with no initial problem or theorem. In Level 1, all attributes of the problem, theorem, or material are listed, and in Level 2, all attributes are negated by asking "what if not attribute x?". In Level 3, the changes that resulted from the negated attributes are formulated in a question. Finally, the posed problems or theorems are reflected upon.

The Thèbault-Yaglom Theorem in section 4.1, although not explicitly mentioned at the beginning, has been investigated in a very similar way to the "What-If-Not" strategy. Figure 4.2 on page 42 graphically illustrates this strategy using the example from section 4.1.

Level 0 As a starting point, we had the Thèbault-Yaglom Theorem.

Level 1 Following "What-If-Not" strategy, we can now list the attributes of the theorem. In Figure 4.2, only some of many possible attributes are listed.

(1) Initial figure is a *parallelogram.*

(2) On each side *squares* are constructed.

(3) Constructed figures on each side are *identical.*

(4) ...

Level 2 For What-If-Not-ing, we concentrated on attribute (1) and asked, what if the initial figure would *not* be a parallelogram? To prove the initial theorem, we even considered a special case first and asked what would happen if the initial figure was not a parallelogram but a square? In the course of section 4.1, we tried to generalize the initial theorem by asking, what if the initial figure was an equilateral triangle, an arbitrary triangle, a regular n-gon or even an affine-regular n-gon. In Figure 4.2, the next level is deepened for the question $(\sim 1)_2$: What if the initial figure were a trapezoid?

Level 3 If we now have a trapezoid as the initial figure, we can ask, for example, $(\sim 1)_2a$ whether the barycenters of the squares constructed on the sides of the trapezoid still form a square. As we have seen, this is not the case. Therefore the question follows, $(\sim 1)_2b$ how the figures constructed on the trapezoid sides must be changed so that the barycenters make a square again.

Level 4 If we now try to analyze these questions and answer those, we notice that we need to make further changes to other attributes from the list from Level 1. We have found that if we want a trapezoid as the initial figure to continue to have the barycenters of the figures constructed on its sides be a square, (~ 2) not only squares and (~ 3) no identical figures must be chosen. Instead, squares had to be constructed on the nonparallel sides of a trapezoid and rectangles had to be constructed on the parallel sides of a trapezoid, so that the heights of the rectangles equals the length of the opposite side of the trapezoid.

In Germany, Schupp (2002) takes up the idea by Brown and Walter. He proposes 24 task variation strategies by means of which students can pursue the activity of task variation in a regulated way. Among these strategies we find *analogize*, *generalize*, *specialize*, but also something like *sensemaking*, *reverse*, or *iterate*.

In subsequent years, researchers in mathematics education have attempted to bring order and structure to this newly conceptualized field. Over the years, different definitions of problem posing have emerged. Table 4.2 on page 43 summarizes some of these widely cited definitions. These definitions always refer to activities of posing problems in the field of mathematics.[3] *Problem* refers in this context to all tasks on the spectrum between routine task and non-routine problem. By non-routine problems, in contrast to routine tasks, we mean tasks for which no procedure for a solution is known and strategies for a solution have to be found and applied (Schoenfeld, 1985b). Article 1 (chapter 6, *Rethinking problem-posing situations: A review*) takes a nuanced look at this spectrum between routine tasks and non-routine problems in the context of problem posing.

The definition of Silver (1994) includes two different activities: *Generating* and *reformulating*. According to Silver (1994), problem posing can occur *before*, *during*, or *after* problem solving. Problem posing occurs before problem solving when the goal is not the solution of a problem but the creation of a new problem. This refers to the *generation* of a new problem. Problem posing occurs during problem solving when someone is stuck in a problem-solving process and tries to make that problem

[3] There is also research in the area of posing researchable questions in mathematics education (Cai & Mamlok-Naaman, 2020; Schoenfeld, 2020), but this is a different field of research.

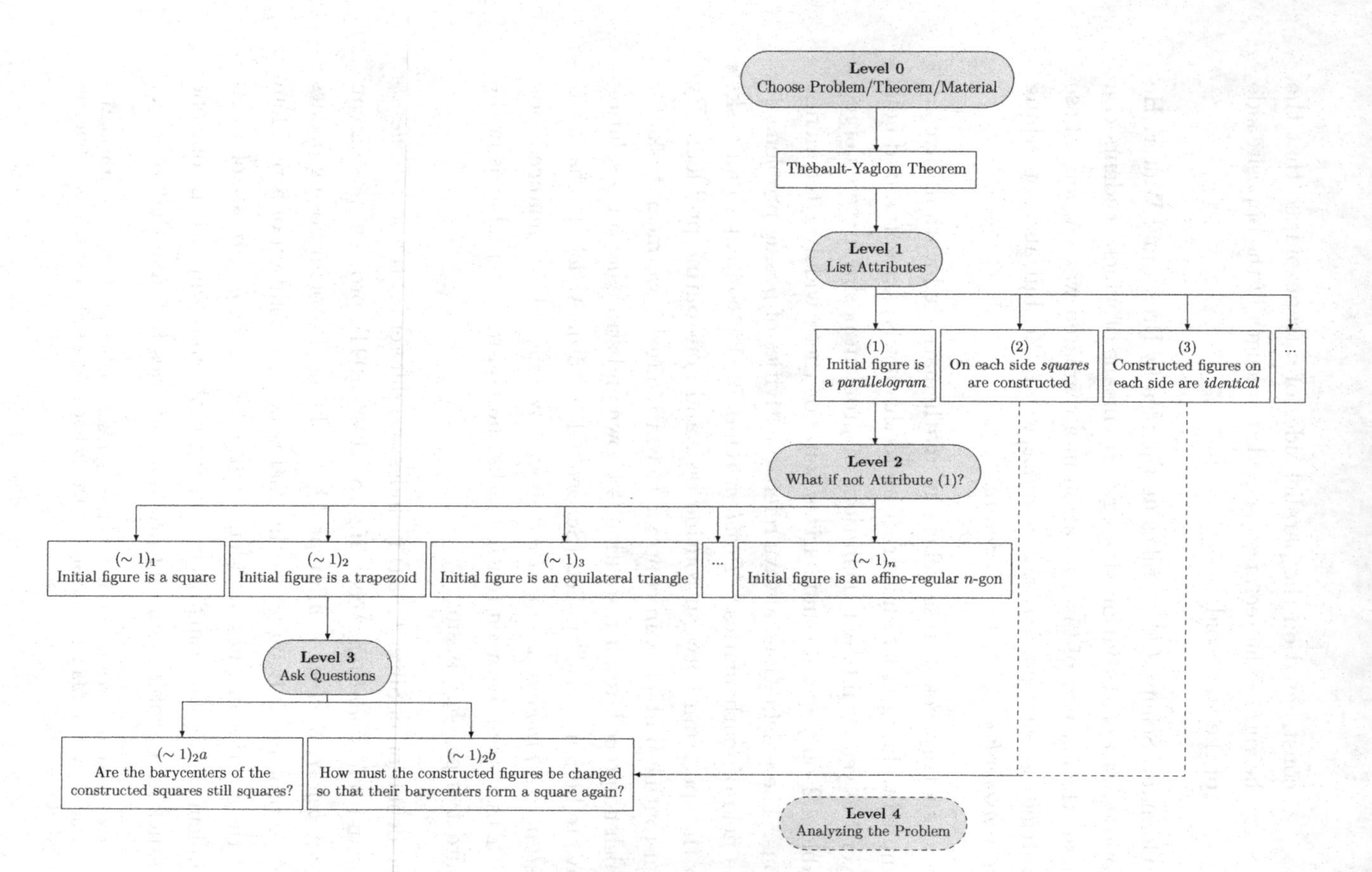
Level 0
Choose Problem/Theorem/Material
Thèbault-Yaglom Theorem
Level 1
List Attributes
(1)
Initial figure is a parallelogram
(2)
On each side squares are constructed
(3)
Constructed figures on each side are identical
...
Level 2
What if not Attribute (1)?
(~ 1)1
Initial figure is a square
(~ 1)2
Initial figure is a trapezoid
(~ 1)3
Initial figure is an equilateral triangle
...
(~ 1)n
Initial figure is an affine-regular n-gon
Level 3
Ask Questions
(~ 1)2a
Are the barycenters of the constructed squares still squares?
(~ 1)2b
How must the constructed figures be changed so that their barycenters form a square again?
Level 4
Analyzing the Problem

Fig. 4.2.: Description of one problem-posing path with Initial problem 1 from section 4.1 using the "What-If-Not" strategy

more accessible by reformulating it. We have seen this kind of problem posing in the beginning of section 4.1 when we had no idea for proving theorem 1 and reformulated it to make it easier. Problem posing can also occur after problem solving. This refers to the activity of investigating a given problem to get more generalized insights into it. This was also carried out in section 4.1 when we posed generalized theorems with regard to the theorem by Thèbault-Yaglom and reached a quite general conjecture due to productive problem posing. Both problem posing during and after problem solving refers to the *reformulation* of a given problem.

Source	Conception
Silver (1994)	"Problem posing refers to both the generation of new problems and the re-formulation, of given problems." (p. 19)
Stoyanova and Ellerton (1996)	"[M]athematical problem posing will be defined as the process by which, on the basis of mathematical experience, students construct personal interpretations of concrete situations and formulate them as meaningful mathematical problems." (p. 518)
Klinshtern et al. (2015)	"[P]roblem posing means an accomplishment that consists of constructing a problem that satisfies the following three conditions: (a) it somehow differs from the problems that appear in the resources available to the teacher; (b) it has not been approached by the students; and (c) it can be used in order to fulfill teaching needs that otherwise could be difficult to fulfill." (p. 463)
Cai and Hwang (2020)	"By problem posing in mathematics education, we refer to several related types of activity that entail or support teachers and students formulating (or reformulating) and expressing a problem or task based on a particular context (which we refer to as the problem context or problem situation)." (p.2)

Tab. 4.2.: Problem-posing definitions

Stoyanova and Ellerton (1996) provide a more student-oriented definition (see Table 4.2 on the previous page). This definition is based on individual mathematical experiences and defines problem posing as a process of interpreting concrete situations. A situation is a not well-structured problem in the sense that the goal cannot be determined by all given elements and relationships (Stoyanova, 1997, p. 5). Stoyanova and Ellerton (1996) distinguish between free, semi-structured, and structured problem-posing situations, depending on their degree of given information (see Table 4.3 on the facing page). Free situations provoke the activity of posing problems out of a given, naturalistic, or constructed situation without any restrictions. In semi-structured situations, posers are invited to explore the structure of an open situation by using mathematical knowledge, skills, and concepts of previous mathematical experiences. In structured situations, people are asked to pose further problems based on a specific problem, for example by varying its conditions. Article 1 (chapter 6, *Rethinking problem-posing situations: A review*) focuses in depth on such problem-posing situations used in studies within a systematic review. Referring to the conceptualization by Koichu (2020), the problem-posing situations in table 4.3 on the next page can be seen as *didactical*, i.e. problem posing is a goal by itself and is explicitly demanded. Besides that, problem posing can be evoked *a-didactically* as an implicit goal. This applies, for example, for open-ended or inquiry-based problem-solving environments (Cifarelli & Cai, 2005; da Ponte & Henriques, 2013; Leikin & Elgrably, 2020). The investigation conducted in section 4.1 was therefore also of an a-didactic nature.

Klinshtern et al. (2015) provide a more teacher-oriented definition. Their definition is the result of a study on what teachers mean by posing problems. For this reason, this definition also includes something like an educational goal of the posed problem.

	Situation
Free	Think of two problems on percents by yourself. Write them down clearly and show solution strategies and answers for both. The first problem should be an easy problem and the second problem should be a difficult one. You are totally free to design your problems. Fire away! (Heuvel-Panhuizen et al., 1995)
Semi-structured	Mr. Miller drew four of the figures in a pattern as shown below. 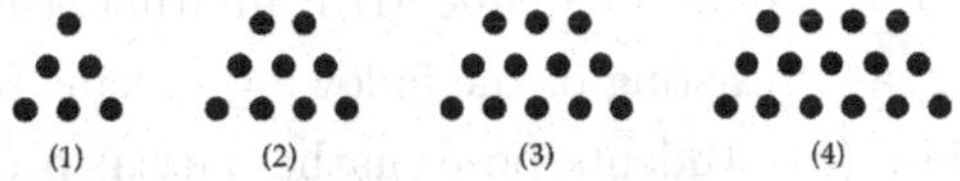 (1) (2) (3) (4) For his students' homework, he wanted to make up some problems according to this pattern. Help Mr. Miller by writing as many problems as you can in the space below. (Cai, 1998)
Structured	[A] sheep is grazing in a square field with side length L. The sheep is tied at the point $(\frac{L}{2}, 0)$, and the rope attached to the sheep has a length R as shown in [the figure below]. 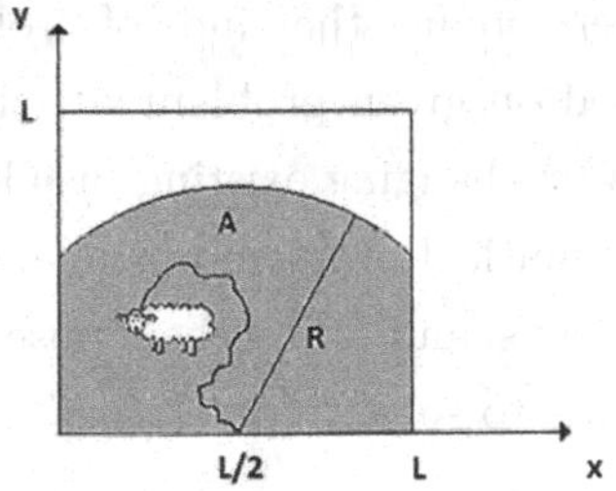 In this figure, A represents the area of the sector where the sheep may graze. Let $r = \frac{R}{L}$ be the ratio of the rope length to field side length, and let $f = \frac{A}{L^2}$ represent the fraction of the total area, which is accessible for the sheep. Obtain f corresponding to one or more values for r. Reformulate the problem in an inverse manner for task enrichment purposes. (cf. Martinez-Luaces et al., 2019a)

Tab. 4.3.: Free, semi-structured, and structured problem-posing situations

Cai and Hwang (2020) integrate student *and* teacher perspectives with their definition. Their conceptualization also includes, in broad terms, the distinction between generating and reformulating in Silver's (1994) definition, as well as the openness of the problem situation inherent in Stoyanova and Ellerton's (1996) definition. Cai and Hwang (2020) specify mathematical problem posing separately for students and teachers as specific intellectual activities:

> "For students, we define MPP [mathematical problem posing] as consisting of the following specific intellectual activities: (a) Students pose mathematical problems based on given problem situations which may include mathematical expressions or diagrams, and (b) students pose problems by changing (i.e., reformulating) existing problems.
>
> For teachers, we define MPP as consisting of the following specific intellectual activities: (a) Teachers themselves pose mathematical problems based on given problem situations which may include mathematical expressions or diagrams, (b) teachers predict the kinds of problems that students can pose based on given problem situations, (c) teachers pose problems by changing existing problems, (d) teachers generate mathematical problem-posing situations for students to pose problems, and (e) teachers pose mathematical problems for students to solve." (p. 3)

With this, Cai and Hwang (2020) make an attempt to represent a wide range of different types of problem posing within the educational field. For students, this leads to the activities of (a) generating and (b) reformulating as described by Silver (1994). Cai and Hwang (2020) also assign these two activities to teachers (in this second definition, generating is labeled as (a) and reformulating is labeled as (c)). In the educational context, according to Cai and Hwang (2020), problem posing for teachers also involves (b) predicting possible problems that students might pose in a given problem-posing situation. This may be due to the fact that when teachers, for example, encourage students to pose problems in class, they are pursuing a certain educational goal, the achievement of which they can anticipate in part by predicting possible posed problems. Moreover, such anticipation helps to be prepared for the versatility of students' posed problems. Cai and Hwang (2020) also describe a type of meta problem posing (d). Teachers also design problem-posing situations for students, that is mathematical contexts that encourage problem posing (Stoyanova, 1997). Lastly, problem posing for teachers involves posing tasks of any kind for students (e).

The articles in Part III are based on the definitions of Silver (1994) and Stoyanova and Ellerton (1996) (see table 4.2 on page 43). Both definitions denote largely equivalent activities. Depending on the study, one of the two definitions were used. The study which will be presented in article 1 (chapter 6, *Rethinking problem-posing situations: A review*) focuses on problem-posing situations. For this reason, the definition of Stoyanova and Ellerton (1996) is used, in which the interpretation of such situation is an integral part of problem posing. The study which will be presented in article 2 (chapter 7, *Developing a framework for characterising problem-posing activities: A review*) focuses on problem-posing activities. For this reason, the definition of Silver (1994) is used, as its differentiation between generating and reformulating helped to characterize different problem-posing activities. The studies which will be presented in articles 3 (chapter 8, *The process of problem posing: Development of a descriptive process model of problem posing*) and 4 (chapter 8, *The process of problem posing: Development of a descriptive process model of problem posing*) emphasize the equivalence of different definitions and integrate the definitions into an overarching problem-posing understanding.

In the following, we want to cover different perspectives on the activity of problem posing to get a more accurate picture of it. In this context, we want to use the *goals* pursued by the respective problem-posing activity as a criterion for differentiation. In the following, we want to cover different perspectives on the activity of problem posing. In this context, we want to use the *goals* pursued by the respective problem-posing activity as a criterion for differentiation. General educational goals related to the activity of problem posing are, among other things, to give an authentic idea of what *doing mathematics* is and to give students a positive self-concept (Schupp, 2002). In this context, however, by goals we do not mean overarching educational goals, but rather focused goals of individual problem-posing sequences. In the following, we will go into more detail about what is meant by this.

To specify what goals we are talking about, we make use of Silver's (1994) conceptualization described earlier, which says problem posing can take place as the generation of new and reformulation of given problems before, during, and after problem solving. In problem posing *before* problem solving, the goal is to create a new problem. *During* problem solving, problem posing aims at making a problem more accessible for a solution. In problem posing *after* problem solving, the goal is to get a more general insight into a mathematical field or to expand the scope of the mathematical content. We want to bring another education-related perspective to problem posing. We already saw this perspective in the conceptions by Klinshtern et al. (2015) and Cai and Hwang (2020). Problem posing can be done not only for oneself, but also for others, that is teachers pose problems for pupils, lecturers for students, or expert problem posers for the contestants of mathematics competitions. This kind of problem posing aims at achieving an educational goal. The goal is to stimulate the construction of new knowledge concerning a mathematical content through appropriately posed problems. These structural considerations gave rise to the following list of different perspectives on problem posing, which is discussed in the following sections:

1. Problem posing as generating new problems
2. Problem posing as reformulating a given problem for problem solving
3. Problem posing as reformulating a given problem for investigation
4. Problem posing as constructing tasks for others

In each section, the corresponding problem-posing activity will first be shown as an example in the form of a mathematical miniature. Then it will be located more generally in the research field by presenting studies that investigate the specific problem-posing activity. The four presented perspectives are by no means separable from one another.

The aim of these four sections is to present a broad spectrum of different problem-posing activities that are conducive for understanding the articles that follow in Part III. Article 1 (see chapter 6, *Rethinking problem-posing situations: A review*) is devoted to the diversity of problem-posing situations. In the following sections, examples are used to indicate the diversity of problem-posing situations. Article 2 (see chapter 7, *Developing a framework for characterising problem-posing activities: A review*) is devoted to the diversity of problem-posing activities. The problem-posing activities described in the following sections are examined in more detail and conceptually distinguished within this article. The two empirical studies presented in Articles 3 (see chapter 8, *The process of problem posing: Development of a descriptive process model of problem posing*) and 4 (see chapter 9, *Identifying metacognitive behavior in problem-posing processes. Development of a framework and a proof of concept*), in particular, elicited problem posing as reformulation among participants. Therefore, sections 4.2.2 and 4.2.3 prepare understanding the empirical studies conducted in articles 3 and

4. Since Articles 3 and 4 are empirical and research related, we will take the opportunity to focus on problem posing in its mathematical side in the coming sections.

4.2.1 Problem posing as generating new problems

Mathematical miniature

To describe the free activity of generating new problems as problem posing, we take a situation where no initial problem is given. Instead, we take an arbitrary material, in this case a 5×5 geoboard (see also Brown & Walter, 2005).

Geoboard
Pose problems for the following 5×5 geoboard.

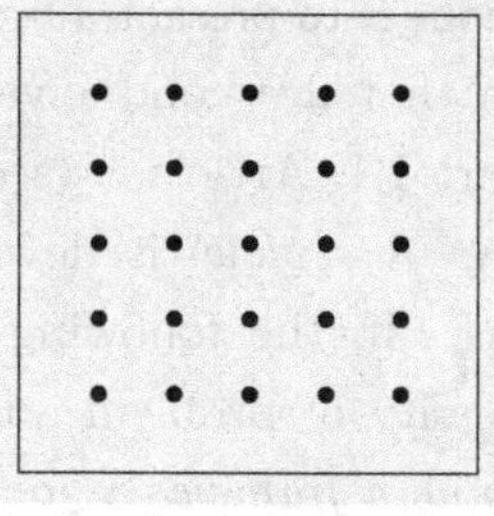

Numerous questions can now be asked about this geoboard. We would like to develop a few ideas in the following. We do not want to pursue a detailed solution to the problems in every case, as the focus is rather on the process of posing. Let us assume that the horizontal and vertical distance of the points has the dimensionless length 1. If you now connect different points on the geoboard, you will find different lengths. Of course all lengths range from 1 to 4. If you connect two diagonally lying points, the length can be determined with the Pythagorean theorem as $\sqrt{1^2 + 1^2} = \sqrt{2}$. This suggests the question of how many different lengths you can find on the geoboard.

> If you connect any two points on the geoboard by a straight line, you will find different lengths. How many different lengths can you find?

Now that we have dealt with different lengths on the geoboard, another obvious question arises because of the square shape of the board. We find different squares on the geoboard. Of course a 1×1 square or the big 4×4 square. But what about other, perhaps non-integer side lengths? Let us first formulate a question about this:

> What is the total number of squares that you can find when connecting dots on a 5×5 geoboard?

We first look at the squares whose sides are parallel to the edge of the geoboard. We find one 4×4 square, four 3×3 squares, nine 2×2 square, and sixteen 1×1 square. These quantities are obviously the square numbers. In addition, we find one square with side length $2\sqrt{2}$, two squares with side length $\sqrt{10}$, eight squares with side length $\sqrt{5}$, and nine squares with side length $\sqrt{2}$. So, in total, there are 50 different squares on a 5×5 geoboard.

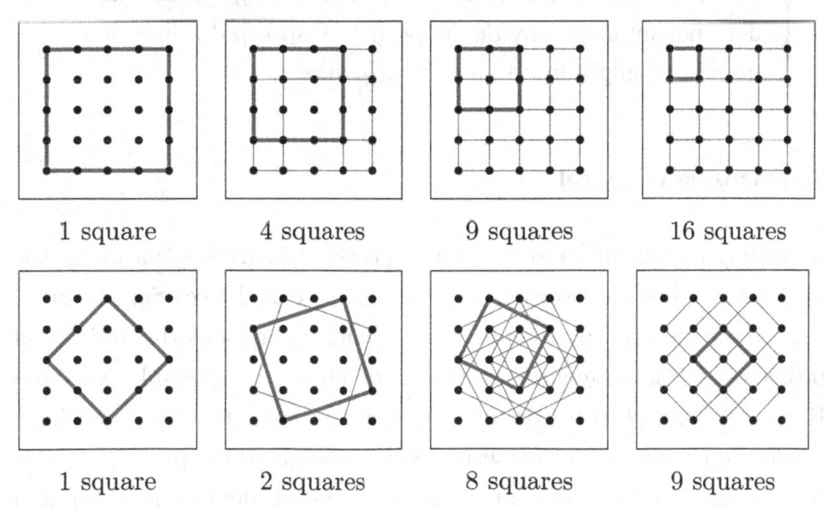

So how can we be sure that we have found all the squares? And if we can answer that, will we be able to generalize these thoughts? Can we identify structures that allow us to determine the number of different squares on a $n \times n$ geoboard? But with such considerations, we are at *Problem posing as reformulating a given problem for investigation* (section 4.2.3) which we want to consider in more detail later.

Now we have dealt more intensively with squares, but what about triangles? Right-angled or isosceles triangles can be found quickly. An equilateral triangle is not so easy to find. So this problem could be formulated:

> Are there equilateral triangles on a 5×5 geoboard?

Now we have posed numerous problems where the dots have been connected. One can quickly find lines within the 5×5 geoboard that do not meet any point. What does this look like when you extend the geoboard to a $n \times n$ geoboard or even a $\infty \times \infty$ geoboard? In this context, there are at least two problems that could be investigated:

> Can I find a line on an unbounded $\infty \times \infty$ geoboard that does not intersect any of the points? Can I find a line that intersects only a finite number of points?

Summarizing empirical research

According to Stoyanova and Ellerton (1996), the given situation of the geoboard without a given initial task can be called free. Some studies use semi-structured problem-posing situations or focus the activity of problem posing as generating new problems. For example, Cai and Hwang (2002) asked participants to pose an easy, medium, and hard problem on a specific dot pattern. They investigated the problem-posing performance of Chinese and US sixth-grade students and compared it to their problem-solving performance. They found a stronger link between problem-posing and problem-solving performance in Chinese students than in US students. Bonotto (2013, see also Bonotto & Santo, 2015) chose a contextualized input. She investigates the potential of so-called artifacts (i.e. real-life objects like restaurant menus, advertisements, or TV guides) to stimulate critical and creative thinking. Bonotto (2013) investigated the relationship between problem-posing and problem-solving activities in 18 primary students and also assessed the use of these artifacts as a semi-structured situation. She concludes that these artifacts from the real world are an appropriate stimulus for problem-posing activities. However, the complexity of the specific artifact, its constraints, and variables should be analyzed before implementing it into the classroom. Silver et al. (1996) had participants exploring a billiard context in their study (see Table 6.6 on page 113, Situation 11) which has also been frequently used in subsequent studies (Kontorovich et al., 2012; Koichu & Kontorovich, 2013). In this context, Koichu and Kontorovich (2013), for example, found that in two successful problem-posing processes when the posers are asked to pose an interesting problem, they also have their own interest in mind. Problem posing then initially consists of searching for an interesting mathematical phenomenon within the given semi-structured situation. In her study, Behrens (2018) uses digital tools to ask 10th and 11th-grade students to pose problems about integer functions. She found that while these students are able to pose such problems, the full potential of the digital tools was not used.

4.2.2 Problem posing as reformulating a given problem for problem solving

Mathematical miniature

As we pointed out above, Pólya (1957) and Schoenfeld (1985b), although they never used the term *problem posing* explicitly, wrote extensively about modifying a problem for solving. Both have in common that for solving a problem they propose to pose a simpler problem. Pólya (1957) describes this in the *Devising a plan* phase. With (Schoenfeld, 1985b), this takes place in the *Exploration* phase. Let's look at an example problem for this.

Example problem

In how many paths can the words *PROBLEM POSING* be read in the figure below?

P
R R
O O O
B B B B
L L L L L
E E E E E E
M M M M M M M
P P P P P P
O O O O O
S S S S
I I I
N N
G

So we are looking for the number of paths leading from the first *P* to the last *G* in this figure. The number of paths to read the words *PROBLEM POSING* does not seem easy to count. For this reason, we modify the task (Pólya, 1957; Schoenfeld, 1985b) and solve a structurally similar, but a simpler task. Let us consider the word *PROBL* first.

> Consider an arbitrary letter in the figure. How many different paths originating at the top (P) lead to this letter?

We will now use the third letter L in the third row to answer this modified task.

$$\begin{array}{ccccc} & & P & & \\ & R & & R & \\ O & & O & & O \\ & B & & B & \\ & & L & & \end{array}$$

To get from the P to on of the R's in the second row of the figure, I obviously have exactly one option in each case. For the O's that are on the very outside of this figure also remain only one option. The O in the middle, however, has two possibilities. From the P, I can go to the right or to the left, and then come to the middle. If I want to go to one of the B's in the line below, I have three choices. Either I go over one of the outer O's or I go one of the two paths over the middle O for I have two options. So slowly a pattern reveals itself. We modify the figure slightly by abstracting it. Instead of the letters, we fill the squares with the number of paths leading from the top to this point.

$$\begin{array}{ccccc} & & 1 & & \\ & 1 & & 1 & \\ 1 & & 2 & & 1 \\ & 3 & & 3 & \\ & & 6 & & \end{array}$$

If we look at this figure, we find that any number other than 1 is the sum of the two adjacent numbers above it. But why is that? As an example, consider again the B in the figure above. If we want to reach B, starting from P, we have to pass through one of the two neighboring O's above it. Once we have reached one of the two O's, there is only one way to get to B from there. So: The total number of ways from P to B is the sum of ways to the two adjacent O's above. Behind this reasoning, of course, there is a general structure that we can also apply to the example problem. To do this, we again replace the letters of the words *PROBLEM POSING* with numbers representing the number of paths that can be walked to this point. The numbers now result from the structure found, namely the sum of the two numbers above.

```
                  1
               1     1
            1     2     1
         1     3     3     1
      1     4     6     4     1
   1     5    10    10     5     1
1     6    15    20    15     6     1
   7    21    35    35    21     7
     28    56    70    56    28
        84   126   126    84
           210   252   210
              462   462
                 924
```

Thus, there are 924 paths to read the words *PROBLEM POSING* in the respective figure. Of course, the interested reader should notice that this structure is a segment of Pascal's triangle. In this exploration, we started with a problem that we could not solve at first. We have made it a structurally similar but simpler problem by modification. We were able to solve it by recognizing a pattern. This pattern could be transferred to the initial problem because of the structural equality so that we could solve this problem as well. So we did *problem posing as reformulating for problem solving.*

Summarizing empirical research

When we look at research on problem posing for problem solving, conceptual work stands out the most. The work of Pólya (1957) and Schoenfeld

(1985b) has already been referred to in this context. In fact, Duncker (1945, pp. 8–9) already pointed out that, for him, problem solving can be conceptualized as productive reformulating of a given initial problem. With regard to empirical research on problem posing as reformulating a given problem for problem solving, we see that in Xie and Masingila (2017) problem posing is quite explicitly intended to be used as a heuristic for solving a problem. They state that "especially posing easier problems than the original one, enhances participants' understanding of the structure of the given problem and further contributed to problem solving" (p. 108). Cifarelli and Sevim (2015) investigate how problem posing contributes to solvers' problem-solving activity and conclude that problem posing should be emphasized by teachers as it is an integral part of problem solving.

4.2.3 Problem posing as reformulating a given problem for investigation

Mathematical miniature

In section 4.1, we first proved the theorem by Thèbault-Yaglom. Subsequently, this theorem was modified by varying selected conditions in order to obtain more general insights. This is already a suitable example of *posing as reformulating given problems for investigation.* However, we want to illustrate this kind of problem posing with a further example from number theory (cf. Ziegenbalg, 2014).

> **Sum of consecutive numbers**
> In how many different ways can you express 15 as the sum of consecutive numbers?

We first consider a possible solution to this task. We notice that we can express 15 as $7 + 8$. Furthermore, let us first express 15 as a sum of equal summands. For this purpose it helps to look at the prime factorization or the divisors of 15. Besides 1, 15 is divisible by 3, 5, and 15. So, we can express 15 as $3 \cdot 5 = 5 + 5 + 5$, $5 \cdot 3 = 3 + 3 + 3 + 3 + 3$, and $15 \cdot 1 = 1 + 1 + 1 + 1 + 1 + 1 + 1 + 1 + 1 + 1 + 1 + 1 + 1 + 1 + 1$. We can cleverly rearrange within these sums, so for example in the sum $5 + 5 + 5$, we decrease the first summand by 1 and increase the last one by 1, so that the sum remains constant. This leads to $3 + 4 + 5$ which is a sum of consecutive numbers. The following figure illustrates this process with figured numbers.

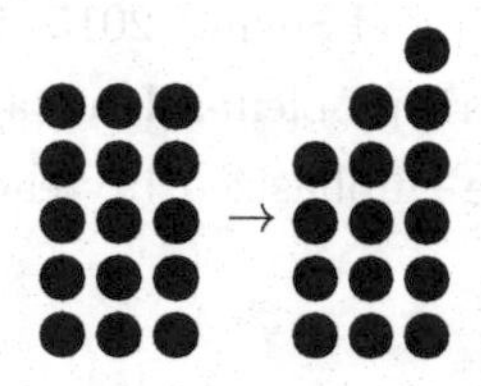

For $3 + 3 + 3 + 3 + 3$ we proceed analogously. The middle 3 remains, the second summand we decrease by 1, the fourth summand we increase by 1. The first summand we decrease by 2, the fifth summand we increase by 2. This leads to the sum $1 + 2 + 3 + 4 + 5$. Also for the sum of 1's, we proceed like that, leading to $(-6) + (-5) + (-4) + (-3) + (-2) + (-1) + 0 + 1 + 2 + 3 + 4 + 5 + 6 + 7 + 8$. The negative integers cancel each other out and the sum $7 + 8$ remains. This leads to:

$$
\begin{aligned}
15 &= 4 + 5 + 6 \\
&= 1 + 2 + 3 + 4 + 5 \\
&= 7 + 8
\end{aligned}
$$

There do not seem to be any other solutions. Interestingly, there seems to be a relationship between the number of divisors and the number of ways to express a number as the sum of consecutive numbers. 15 has three divisors not equal to 1 and can be expressed in three ways by the sum of consecutive numbers. Let us now begin to get to a more general insight in the sense of *investigating through reformulating* this initial problem. For that, we pose a similar problem in which we consider a different number which we want to express as a sum of consecutive numbers with the same approach:

> In how many different ways can you express 20 as the sum of consecutive numbers?

20 is, besides 1, divisible by 2, 4, 5, 10, and 20. So, we can express 20 as $10+10$, $5+5+5+5$, $4+4+4+4+4$, $2+2+2+2+2+2+2+2+2+2$, and $1+1+1+1+1+1+1+1+1+1+1+1+1+1+1+1+1+1+1+1$. If we now want to make the same rearrangement as with the above sums of the number 15, we find that we do not succeed in every case. $10+10$ cannot be rearranged in such a way, since we have an even number of summands, with which we cannot rearrange with constant sum. The same applies for the sum $5+5+5+5$, $2+2+2+2+2+2+2+2+2+2$, and $1+1+1+1+1+1+1+1+1+1+1+1+1+1+1+1+1+1+1+1$. Only in the case of $4+4+4+4+4$ can we perform our procedure and arrive at the only solution of expressing 20 as the sum of consecutive numbers.

$$20 = 2+3+4+5+6$$

This means that not only the number of divisors plays a role, but also their parity. The number 15 has three odd divisors grater than 1, and consequently it can be written as a sum of consecutive numbers in three different ways. The number 20 has only one odd divisor above 1 and therefore one expression as the sum of consecutive numbers. If this pattern continues, there should be no expression as a sum of consecutive numbers for numbers that have only even divisors, for example, $16 = 2^4$. So let's pose the next problem:

> In how many different ways can you express 16 as the sum of consecutive numbers?

16 has, besides 1, the divisors 2, 4, 8, and 16 which leads to the sums $8+8$, $4+4+4+4$, $2+2+2+2+2+2+2+2$, and $1+1+1+1+1+1+1+1+1+1+1+1+1+1+1+1$. According to our expectation, these sums cannot be cleverly rearrange, so that a sum of consecutive numbers results.

Our assumption is now solidified and we can state: Any natural number can be represented as a sum of consecutive numbers in as many ways as it has odd divisors > 1. This is actually called Sylvester's Theorem after the British mathematician James Joseph Sylvester.

Summarizing empirical research

This adjacency of problem posing and problem solving suggested in this section is also explored in the context of an open problem situation by Cifarelli and Cai (2005). They observed two students solving a problem related to billiard (similar to situation 11 in Table 6.6 on page 113) and found that problem posing and problem solving were always intertwined. They state that "problem posing and solving appeared to evolve simultaneously, each informing and serving as a catalyst for the other as solution activity progressed" (p. 321). They made similar observations in a later study (Cifarelli & Sevim, 2015). da Ponte and Henriques (2013) examine the problem-posing process in investigation tasks among university students and found that problem posing and problem solving complement each other in generalizing or specifying conjectures to obtain more general knowledge about the mathematics contents. Voica and Singer (2013) analyzed the products of 42 students with above-average mathematical abilities modifying a given initial problem. They found evidence of links between the quality of the students' posed problems and their cognitive flexibility. Martinez-Luaces et al. (2019b) also used a structured situation to encourage prospective teachers to enrich an initial problem by modifying it (see Table 4.3 on page 45). They found out that the prospective teachers have been very creative in the reformulation of the given problem as well as the enrichment of the problem itself. However, some of the participants trivialized the initial problem. As Martinez-Luaces et al. (2019a) states, this should not be criticized as these trivializations sometimes provide further insights into the problem itself.

4.2.4 Problem posing as constructing tasks for others

Mathematical miniature

In this section, we will look at what problem posing can look like when you do it for others rather than for yourself. As an example, we use a specific topic from number theory. A typical task in number theory is to determine all divisors of a number. Using the prime factorization of a natural number $n > 1$, one can arrange the divisors of n in a *Hasse diagram* representing the divisibility relation in the divisor set T_n. Textbooks on elementary number theory often have exercises that involve Hasse diagrams (Padberg, 2008). In a Hasse diagram, there is a connection from $t_1 \in T_n$ to $t_2 \in T_n$ if and only if $t_1 \mid t_2, t_1 < t_2$ and there is no $s \in T_n$ with $t_1 \mid s$ and $s \mid t_2$. In the following figure, the Hasse diagrams are given for $8 = 2^4$ and $12 = 2^2 \cdot 3$.

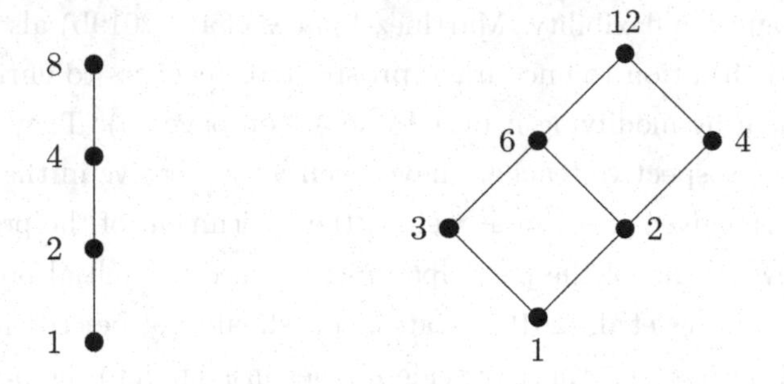

As a hypothetical lecturer in a number theory course, I now want students to use the Hasse diagrams to represent the divisors of a number in a structured way. To understand the structure of Hasse diagrams, students should first draw divisor sets of certain numbers and their corresponding Hasse diagrams. To make the Hasse diagrams different from each other, I choose the task so that the number of prime factors of the numbers differs. This applies, for example, to the numbers $81 = 3^4$, $104 = 2^3 \cdot 13$, and $350 = 2 \cdot 5^2 \cdot 7$. Therefore, I first pose the following task for the students:

> Determine the divisor sets of the numbers 81, 104, and 350. Draw the corresponding Hasse diagrams.

This leads to the following solution:

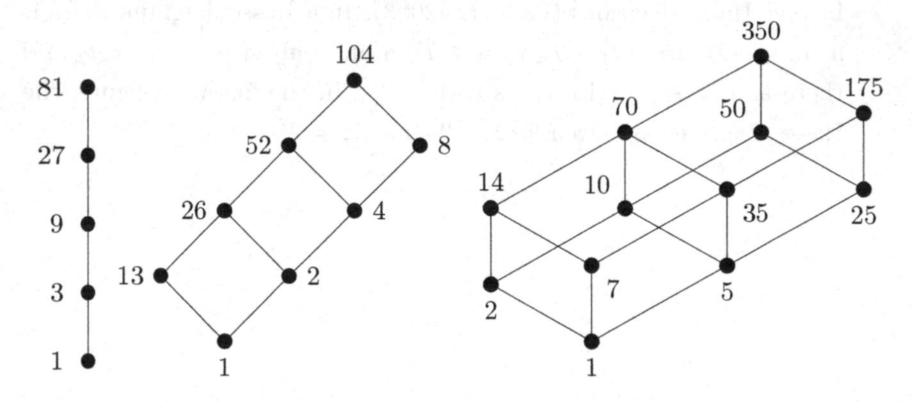

It now occurs to us that for a number with four prime factors (e.g., $9828 = 2^2 \cdot 3^3 \cdot 7 \cdot 13$), we would have to draw a four-dimensional Hasse diagram. How could this look like? Students could come across this insight with the following task:

> Try to draw the Hasse diagram of the number 9828. Why is this difficult? Can you generalize these thoughts?

Now, to encourage a more flexible use of Hasse diagrams, I reverse the previous task on one side and generalize it[4]:

> Describe the set of all numbers whose Hasse diagrams have the following form:

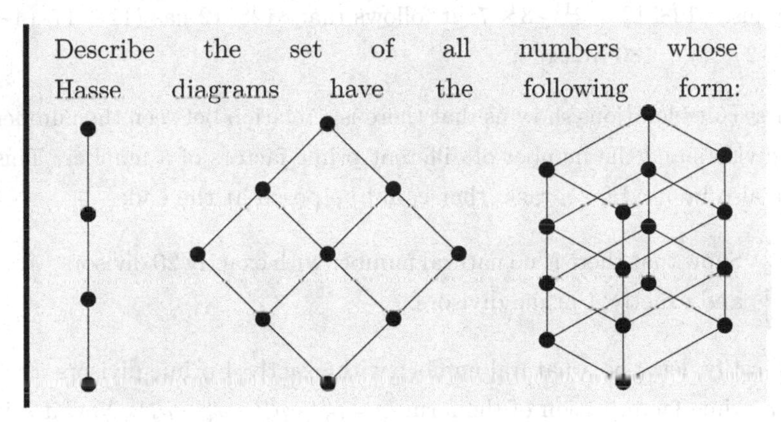

From left to right, this leads to the following statements

- All numbers $n \in \mathbb{N}$ whose prime factorization is of the form $n = p^3$ with any prime p.
- All numbers $n \in \mathbb{N}$ whose prime factorization is of the form $n = p_1^2 \cdot p_2^2$ with any primes p_1, p_2.
- All numbers $n \in \mathbb{N}$ whose prime factorization is of the form $n = p_1 \cdot p_2^2 \cdot p_3^3$ with any primes p_1, p_2, p_3.

[4] Note that two strategies of task variation according to Schupp (2002) were used here, namely *invert* and *generalize.*

From the Hasse diagrams we get the following relation: A natural number n with the prime factorization $n = p_1^{n_1} \cdot p_2^{n_2} \cdot \ldots \cdot p_k^{n_k}$ has $(n_1 + 1)(n_2 + 1) \cdot \ldots \cdot (n_k + 1)$ divisors. Accordingly, the number of divisors can be read directly from the prime factorization. This should also find its way into a posed task to the students.

> The Hasse diagram shows all divisors of a specific number. How does this help you to determine the number of divisors of 8128512?

Because $8128512 = 2^{11} \cdot 3^4 \cdot 7^2$ it follows that 8128512 has $(11+1) \cdot (4+1) \cdot (2+1) = 180$ divisors.

These considerations show us that there is a relation between the number of divisors and the number of different prime factors of a number. This can also be used for a task, that could be posed at the end:

> Show that there is no natural number with exactly 20 divisors and exactly 4 prime divisors.

To justify, let k be a natural number with exactly 4 prime divisors, that is, a prime factorization of the form $k = p_1^{n_1} \cdot p_2^{n_2} \cdot p_3^{n_3} \cdot p_4^{n_4}$. Now if k is to have exactly 20 divisors, $(n_1 + 1)(n_2 + 1)(n_3 + 1)(n_4 + 1) = 20$ must be true. 20 has the prime factorization $20 = 2^2 \cdot 5$. Thus, the number 20 cannot be a product of four integers greater than one. This leads to a contradiction, which shows that there is no natural number with exactly 20 divisors.

Summarizing empirical research

Looking at research done on *problem posing as constructing tasks for others*, we see several studies that investigate in- or pre-service teachers that are asked to pose problems for their students based on a specific given task or content. In this context, problem posing fulfills an educational goal because, for example, teachers want to achieve a certain learning goal with the students using tasks posed by the teacher (cf. Cruz, 2006). Problem posing can also be a diagnostic tool, because thoughtful posed problems can reveal students' definitions and misconceptions (Chen et al., 2011; Tichá & Hošpesová, 2013). For example, Nicol and Crespo (2006) have found that two preservice teachers in their study expanded the mathematical content of the selected textbook tasks to make them more complex. The preservice teachers were also asked to collect tasks for teaching based on the available textbooks. In this context, they indicated a preference for self-posed problems.

Textbook authors also need to set numerous tasks of different kinds (e.g., exercises, contextualized tasks, proving tasks) on different topics. This *constructing tasks for others* also includes, of course, the creation of tasks for mathematics competitions (Poulos, 2017; Kontorovich, 2020; Kontorovich & Koichu, 2016; Sharygin, 2001). For example, Kontorovich and Koichu (2016) observed an expert problem poser posing problems for mathematics competition. They have found that the expert problem poser draws new problems from a family of familiar problems. Moreover, the expert problem poser was concerned that the solution is perceived as novel and surprising not only to potential solvers but also to himself.

5 Problem posing and its connection to related constructs in mathematics education

Now that we have established problem posing as an activity unto itself, we would like to locate problem posing in general in its neighborhood among adjacent activities and constructs. In particular, we want to look at three activities or constructs:

1. **Modelling:** In mathematical modeling, real-world situations are mathematized. Since asking questions arises in a very natural way within modelling, problem posing can also be conceptualized as modeling.

2. **Proving:** In section 4.1 on page 21, we saw how closely problem posing and proving are intertwined in the context of the theorem by Thèbault-Yaglom. In this chapter, we want to further differentiate this connection.

3. **Problem Solving:** The proximity between problem posing and problem solving has already been suggested above. We want to take the opportunity to highlight different perspectives between the relationship between problem posing and problem solving.

4. **Creativity and giftedness:** Like problem solving, problem posing is often associated with creativity and giftedness. We

L. Baumanns, *Mathematical Problem Posing*, Kölner Beiträge zur
Didaktik der Mathematik, https://doi.org/10.1007/978-3-658-39917-7_5

will conclude this theoretical section by situating problem posing within *mathematical* creativity and giftedness research as well as within *general* creativity and giftedness research.

Figure 5.1 summarizes the structure of this section.

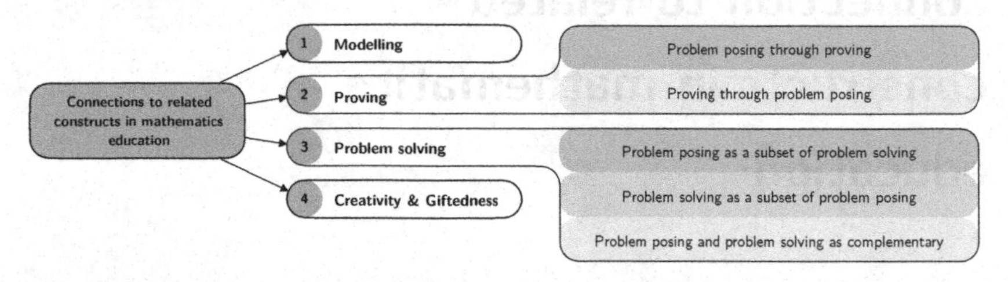

Fig. 5.1.: Structure of this section

5.1 Modelling

Mathematical modeling includes several activities by means of which real-world problems are translated into the world of mathematics in order to deal with it using the tools of mathematics. According to Blum and Leiß (2007), these activities are *understanding* a real-world problem, and *simplifying* it to a real-world model. Subsequently, this model is *mathematized.* The resulting mathematical model is then *worked on.* Finally, the results are *interpreted, validated* and *exposed.* The modelling cycle according to Blum and Leiß (2007) is illustrated in figure 5.2 on the next page.

The perspective of seeing problem posing as modeling is rarely taken (Papadopoulos et al., 2022). Bonotto (2010) writes based on the problem-posing definition by Stoyanova and Ellerton (1996; see Table 4.2 on page 43): "This process is similar to situations to be mathematised, which students have encountered or will encounter outside school." (p. 402) English et al. (2005) take a similar perspective, but reverse it. They

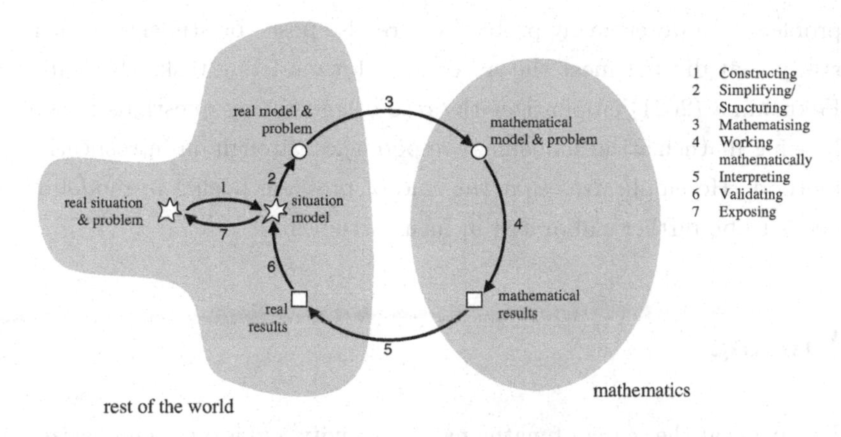

Fig. 5.2.: Modelling cycle according to Blum and Leiß (2007)

highlight the extent to which modelling is an activity consisting of problem posing and problem solving. More specifically, English (2020) says:

> "Modeling problems naturally encourage problem posing as students work collaboratively in model generation. Numerous problems, questions, issues, conflicts, revisions, and resolutions arise as students develop, assess, and prepare to communicate their models to their peers"

Also, Greer (1992) describes problem posing as an activity of "translating from the natural language representation of a problem to the mathematical-language representation of the model" (p. 285). Stillman (2015) says that problem posing in a real-world situation occurs when a problem is formulated in such a way that it can be subjected to mathematical analysis. Thus, there are numerous situations in the surrounding world that can be transformed into a solvable problem.

Regarding a connection between problem posing and modeling, Hartmann et al. (2021) state that students are capable of posing modeling problems. However, many problems were also posed by students in their study that did not meet the set criteria for modeling tasks. Recently, Fukushima (2021) emphasizes the role of generating questions for enhancing mathematical modeling competencies through an introspective method. He emphasizes that the role of problem posing in modeling needs to be further elaborated in future studies.

5.2 Proving

Proving is at the center of mathematical activity and is often emphasized as a unique feature of mathematics that distinguishes it from other (natural) sciences. Perspectives on proofs are as diverse as the proofs themselves. With his *Elements*, Euclid has significantly shaped the image of proofs in mathematics. He proves his theorems by means of a few definitions, postulates, and axioms. With this ideal of mathematics, which is still influential today, proofs can be seen as the justification of a theorem from axioms of a theory by means of deductive reasoning (Meschkowski, 1990).

There are several typical proof procedures that are outlined in the following:

- **Direct proof:** Taking premises (A) or theorems ($A_1, \ldots, A_n$) that have already been proven to be true and deriving, by logical deduction, from them the proposition (B) to be proved.
  $A \Rightarrow A_1 \Rightarrow A_2 \Rightarrow \ldots \Rightarrow A_n \Rightarrow B$

- **Proof by contradiction:** Show that a contradiction arises if the proposition to be proved (B) were false.
  $(\neg B \Rightarrow \lightning) \Rightarrow B$

- **Proof by contrapositive:** Instead of the statement of the form "A implies B", the logically equivalent statement "not B implies

not A" is proved.
$\neg B \Rightarrow \neg A \Leftrightarrow A \Rightarrow B$

- **Pigeonhole principle:** If $n+1$ objects are distributed over n drawers, at least one drawer contains at least two objects.

- **Induction:** If a statement A is to be proved for all natural numbers $\mathbb{N}$ greater than or equal to a certain starting value n_0, show the statement applies for the starting element and then conclude that if the statement holds for a number $n \in \mathbb{N}$, it also holds for $n+1$. Then the statement holds for all $n \in \mathbb{N}_{\geq n_0}$.
$A(n_0) \wedge \Big(A(n) \Rightarrow A(n+1)\Big) \Rightarrow \forall n \in \mathbb{N}_{\geq n_0} : A(n)$.

Proofs have different purposes. Hersh (1993) says, the purpose of proofs in mathematics research is to convince, but in the classroom the purpose of proofs is to explain. The following list collects several other and further purposes of proofs (Hanna, 2000; de Villiers, 1990):

- verification (concerned with the truth of the statement)
- explanation (providing insight into why it is true)
- systematization (the organization of various results into a deductive system of axioms, major concepts, and theorems)
- discovery (the discovery or invention of new results)
- communication (the transmission of mathematical knowledge)
- construction (of an empirical theory)
- intellectual challenge (the self-realization/fulfillment derived from contructing a proof)
- incorporation (of a well-known fact into a new framework and thus viewing it from a fresh perspective)

With regard to the relationship between problem posing and proving, we do not intend to show conceptual similarities between the two activities here. Instead, we want to emphasize two perspectives on how the two activities complement each other: (1) problem posing through proving and (2) proving through problem posing.

5.2.1 Proving through problem posing

In section 4.1 on page 21, a goal was to prove the theorem by Thèbault-Yaglom. To achieve this goal, we first posed and proved a special case of this theorem. This helped prove the original theorem. In this sense, we have been doing problem posing as proving through. This aspect of the interplay between problem posing and proving is similar to the interplay between problem posing and problem solving. Problem posing can also help in problem solving by posing a simpler problem, the solution of which could help in solving the actual problem. This aspect will be discussed in more detail in chapter 5.3 on the facing page.

5.2.2 Problem posing through proving

When one has proved a theorem, a subsequent theorem often arises immediately, where one may be able to apply the findings of one proof to the next. This is what occurred in section 4.1 on page 21, where ideas for subsequent theorems emerged from the proof of the theorem by Thèbault-Yaglom. In this sense, we did *problem posing through proving.* However, the proofs in the theorem by Thèbault-Yaglom were not structurally equivalent to the proofs of the generalizations.

Leikin and colleagues investigated problem posing through proving (Leikin & Grossman, 2013; Leikin, 2015; Leikin & Elgrably, 2020). For example, Leikin (2015) conceptualized the interwoven interplay between problem posing, proving, and investigation in the context of geometry and dynamic geometry environments in a study with prospective secondary school mathematics teachers.

5.3 Problem Solving

In mathematics education, the term problem has for a long time had different meanings depending on the context, which makes its use sometimes diffuse (Schoenfeld, 1992; Lester, 1980). Nowadays, the term *problem* is understood as a certain type of task in which no procedure for solving the task is known at the moment of dealing with it (Schoenfeld, 1985b). Typical activities in mathematical problem solving are therefore, for example, searching for and applying strategies. Among the various conceptions describing the problem-solving process (Artzt & Armour-Thomas, 1992; Fernandez et al., 1994; Mason et al., 2010; Rott et al., 2021), those of Pólya (1957) (see figure 5.3 on the following page) and Schoenfeld (1985b) (see figure 5.4 on the next page) will be outlined.

In the model by Pólya (1957), in the first phase of *understanding the problem*, students should be able to understand the wording of the problem and repeat it with their own formulations. By asking adequate questions about known and unknown information as well as the given conditions of the problem, the teacher can support this understanding. The second phase is about *developing a plan.* Pólya states that related problems or dividing the actual problem into subproblems can help to find the solution of the actual problem. In the third phase, the devised plan is *carried out.* In the final *look back* phase, the solution path and the plausibility of the solution are reviewed, further possible solutions are sought, and the applicability of the solution path to other tasks is considered.

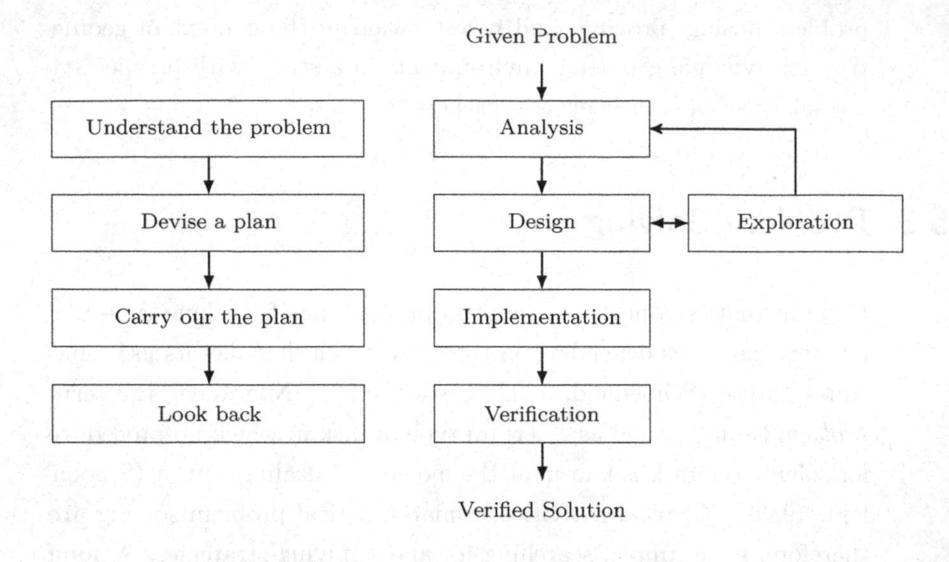

Fig. 5.3.: Pólya's (1957) phase model of problem-solving processes

Fig. 5.4.: Schoenfeld's (1985b) phase model of problem-solving processes

The model by Schoenfeld (1985b) implicitly builds on the model by Pólya (1957). After the first phase of *analysis* of the problem, which largely resembles Pólya's first phase, a plan is *designed.* This planning is not understood as a separate box of the process, but as a superordinate meta-process in which self-regulatory mechanisms take effect and the problem-solver is supposed to keep track of his/her own activities. The heuristic heart lies in the *exploration,* to which one arrives at when difficulties arise. Under this Schoenfeld also includes task modifications that are intended to favor problem solving. Then, analogously to Pólya, follows the *implementation* of the designed plan and finally the *verification* of the results.

For problem posing, the term *problem* is still used for any kind of task along the spectrum between routine task and problem (Baumanns &

Rott, 2021; Cai & Hwang, 2020). That problem posing and problem solving are seen as similar or related activities is indicated, among other things, by a edited book with the titles *Posing and solving mathematical problems. Advances and new perspectives* (Felmer et al., 2016) as well as a Special Issue in *ZDM – Mathematics Education* with the title *Empirical Research on Mathematical Problem Solving and Problem Posing Around the World: Summing Up the State of the Art* (Liljedahl & Cai, 2021).

With regard to the empirical relationship between problem posing and the process of problem solving, we find a plenty of studies investigating the relationship between problem-posing and problem-solving abilities (Bonotto & Santo, 2015; Cai, 1998; Cai & Hwang, 2002; Abu-Elwan, 2002; Arıkan & Ünal, 2015; Chen et al., 2011; Chen et al., 2015; Christou, Mousoulides, Pittalis, & Pitta-Pantazi, 2005; Kar et al., 2010; Kılıç, 2017; Kopparla et al., 2018; Priest, 2009; Rosli et al., 2013; Rosli et al., 2015; Sayed, 2002; Silver & Cai, 1996; Walter & Brown, 1977; Xie & Masingila, 2017). Broadly speaking, the finding of numerous studies is that there is a strong relationship between problem-posing and problem-solving abilities.

With regard to the conceptual relationship between problem posing and the problem solving, Dickman (2014, p. 48) states that "[m]athematical problem posing has always been intimately related to mathematical problem solving". Cai et al. (2015) even make the relationship between problem-posing and problem-solving skills to one of the main open research questions in problem-posing research. In fact, drawing on Ruthven (2020), "it is hard to draw a clear line of demarcation between them" (p. 6). With regard to the conceptual location of these two activities, we want to adopt at least three perspectives taken within the research literature: (1) problem posing as a subset of problem solving ($PP \subseteq PS$) , (2) problem solving as a subset of problem posing ($PS \subseteq PP$), and (3) problem posing and problem solving as complementary, but different in nature ($PP \cap PS \notin \{\emptyset, PP, PS\}$) (see figure 5.5).

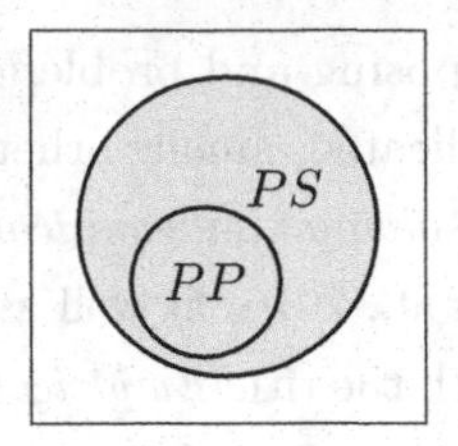

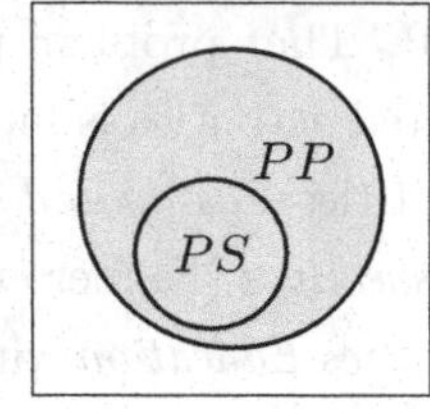

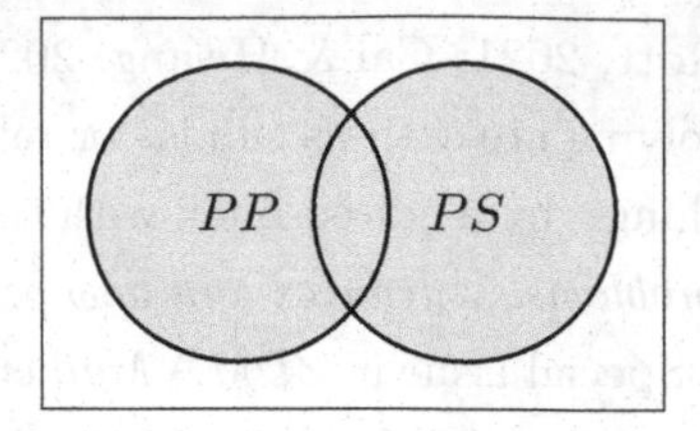

Fig. 5.5.: Perspectives on the relationship between problem posing and problem solving

5.3.1 Problem posing as a subset of problem solving

There is a widespread perspective that problem posing is a special case of problem solving (Silver, 1995; Kontorovich et al., 2012; Arıkan & Ünal, 2015). For example, Silver (1995) writes:

> "The process of problem posing can be considered as a problem-solving process in which the solution (that is, a problem to be posed) is ill-defined, since there are many problems that could be posed. Thus, mathematical problem posing is itself a type of open problem." (p. 69)

A similar perspective is taken by Pehkonen (1995), who conceptualizes problem posing as a problem, for which the starting situation and the goal situation are open. Problem variations, which we have also identified as problem posing, is conceptualized by Pehkonen (1995) as a problem in which either the starting situation or the goal situation is open. For example, in the example from section 4.1 on page 21, the starting point, that is the given theorem, on the one hand, is closed, which means exactly explained. The goal situation, on the other hand, is open as no path of exploration is specified.

Kontorovich et al. (2012) also take this perspective with regard to the relationship between problem posing and problem solving:

> "[I]t is reasonable to consider problem posing as a special case of problem solving. [...] [A] given state is a problem-posing task, i.e., some set of conditions and instructions from which problem-posing process is to begin. A góal state is formulation of a new, at least for the posers, problem that satisfies the requirements of the problem-posing task. Finally, the methods of formulating new problems are, as a rule, not known in advance to problem posers." (p. 151)

In both the quote from Silver (1995) and Kontorovich et al. (2012), problem posing itself is seen as an activity of problem solving. Another perspective of this kind is that problem posing is *part* of the problem-solving process. Fernandez et al. (1994) propose a phase model for problem solving, closely related to the seminal work by Pólya (1957) (see figure 5.6 on the following page). They arrange the four phases by Pólya cyclically. As a transition between these phases, they describe managerial processes, that is self-regulatory activities. Problem posing in this model is entry or exit from problem solving. Thus, their model incorporates parts of Silver's (1994) problem-posing conception presented above, namely that problem posing can take place *before* and *after* problem solving. They do not consider problem posing during problem solving as part of the "making a plan" phase. A very similar approach is taken by Leung (2013) (see figure 5.7 on page 79). She also arranges the Pólya's phases cyclically. Problem posing, however, takes place instead of understanding the problem. If one has posed a problem oneself, understanding does not necessarily take place.

In the context of their study on problem posing within problem solving as reformulation, Cifarelli and Sevim (2015) also summarize that "problem posing has to be seen as integral to the problem-solving process and needs to be emphasized by mathematics teachers at all levels accordingly" (p. 192)

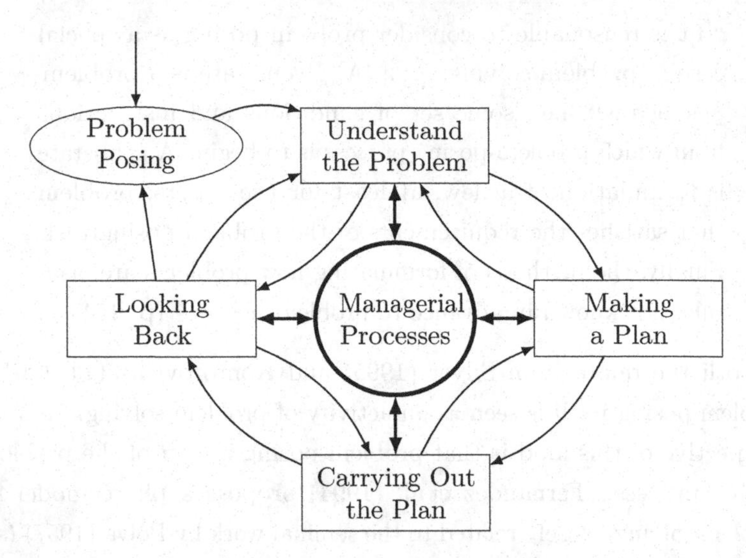

Fig. 5.6.: Phase model for problem solving by Fernandez et al. (1994, p. 196)

This relation between problem posing and problem solving is closely related to the activity of *Problem posing as reformulating a given problem for problem solving* presented in section 4.2.2 on page 54.

5.3.2 Problem solving as a subset of problem posing

Much less often is the perspective taken that problem solving is a subset of problem posing. Kilpatrick (2016) refers in this regard to the statements of Duncker (1945) and Pólya (1957). We have already pointed out that Duncker (1945, pp. 8–9) states, problem solving can be conceptualized as productive reformulating of a given initial problem. Pólya (1957) says in his seminal work on problem solving:

> "We often have to try various modifications of the problem. We have to vary, to restate, to transform it again and again till we succeed eventually in finding something useful. We

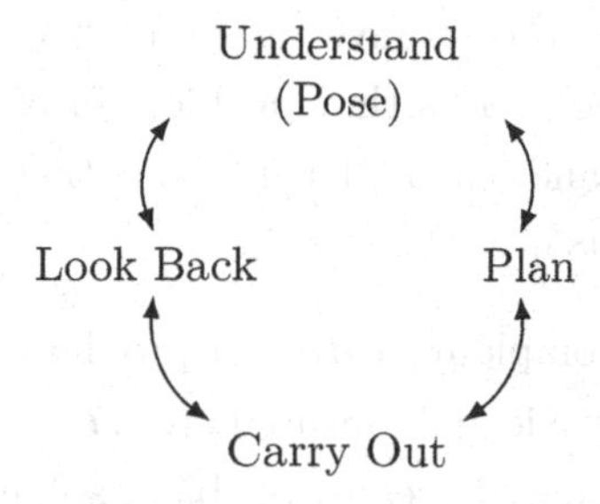

Fig. 5.7.: Phases in problem posing and problem solving by Leung (2013, p. 105)

> may learn by failure; there may be some good idea in an unsuccessful trial, and we may arrive at a more successful trial by *modifying* an unsuccessful one. What we attain after various trials is very often ... a more accessible auxiliary problem." (p. 213)

Thus, both Duncker and Pólya recognize that successful problem solving seems to necessarily involve reformulating. However, outside of these conceptual descriptions and suggestions, we hardly find any empirical studies (e.g., on problem solving) that take this perspective.

5.3.3 Problem posing and problem solving as complementary, but different in nature

Finally, problem posing and problem solving can bee seen as related mathematical activities that complement each other but are different in nature (Dickman, 2014; Cifarelli & Sevim, 2015; Pelczer & Gamboa, 2009). Silver (1994) already separates problem posing and problem solving as activities through his prominent conceptualization described above. Davis (1985) also writes with regard to investigations in mathematics that problem posing and problem solving are parallel activities that complement each other in investigations: "What typically happens in a prolonged investigation is that problem formulation and problem solution so hand in hand, each eliciting the other as the investigation progresses" (Davis, 1985, p. 23).

This adjacency and complementarity of problem posing and problem solving is also noted by Xie and Masingila (2017). On the one hand, they say, problem posing contributes to problem solving. For example, as we also have pointed out, they observed that participants use solutions of simpler problems that they posed, for solving the actual problem. On the other hand, they say that problem solving supports problem posing. They observed that solving a problem helped the participants posing new ones. As problem posing seems to be a quite unfamiliar activity for most of their participants, solving problems first helped them generating some ideas to get the problem posing started. Pelczer and Gamboa (2009) justify separating problem posing and problem solving from another with their empirical studies, in which they observed that "sub-processes involved in the stages [of problem posing] are different and the knowledge has to be applied in a distinct way" (p. 359) compared to problem solving. Cifarelli and Sevim (2015) also investigated this dynamic partnership of problem posing and problem solving. They observed that the "coevolving processes of problem solving and posing" (p. 190) helped of of their participants reaching from a low level of generalization to higher levels that helped getting more general insights into the specific mathematical content.

This relation between problem posing and problem solving is closely related to the activity of *Problem posing as reformulating a given problem for investigation* presented in section 4.2.3 on page 57.

The articles in Part III of this thesis follow the perspective presented in this section, problem posing and problem solving are complementary, but different in nature. It follows that aspects of past problem solving research are taken, but only as an impetus to examine the peculiarities of the process of problem posing.

We have noted in this chapter that the proximity between problem posing and problem solving seems to be so close, theoretically and empirically, that, with recourse to the house metaphor at the beginning of the Theoretical Frame, one could rather speak of semi-detached halves of the same house for the problem-posing house and problem-solving house.

5.4 Creativity and Giftedness

As we stated above, problem posing is often seen as an act of creation or invention. For this reason, problem posing is often named together with creativity in general and mathematical creativity specifically. We will first have a look at the relation between problem posing and creativity in general. Afterwards, we will go into more detail on the relationship between problem posing and mathematical creativity in particular.

Psychological research on problem finding originates mostly on Wallas (1926). Wallas (1926) provided a model of creative thinking consisting of the stages *preparation*, *incubation*, *illumination*, and *verification*. *Preparation* describes thoroughly understanding the problem, *incubation* means the unconscious or automatic approach solving a problem, *illumination* ist the Aha! moment when internally generating an idea after the unconscious incubation, and the final *verification* describes approaches to determine whether the idea is purposeful. Wallas (1926) makes more anecdotic reference to Hermann von Helmholtz and Henri Poincaré with regard to these four phases of the creative process. As Runco and Nemiro (1994) states, "[t]he preparation stage obviously parallels what is now usually called *problem finding*" (accentuation by the author). *Problem finding* is the term, we often find in publications in the environment of psychological publications. Getzels and Csikszentmihalyi (1975) build on Wallas's ideas and investigated the relation between problem-finding performance and quality of the final creative product of art students (Getzels & Csikszentmihalyi, 1975; Csikszentmihalyi & Getzels, 1971). They found out that there is a strong relationship between specific problem-finding variables (e.g., exploration) before the actual drawing and the overall aesthetic value and the originality of the final drawings. Jay and Perkins (1997) identify problem posing – which they refer to as *problem finding* – as a key aspect of creative thinking in general.

With regard to *mathematical* creativity, Silver (1997) has laid the conceptual groundwork for numerous studies (Yuan & Sriraman, 2011; Gleich, 2019; Van Harpen & Sriraman, 2013; Singer & Voica, 2015; Nuha et al., 2018; Bicer et al., 2020; Bonotto & Santo, 2015; Haylock, 1997; Leikin & Elgrably, 2020; Leung, 1993b; Leung, 1997; Pelczer & Rodríguez, 2011; Singer et al., 2011; Singer et al., 2017; Singer & Voica, 2017; Voica & Singer, 2013; Yuan & Sriraman, 2011). Already in 1994, Silver wrote in this regard: "Since problem posing has been embedded in the assessment of creativity or mathematical talent, it is reasonable to assume that there is some link between posing and creativity" (p. 21). As he states, the connection between problem posing and creativity remais uncertain to that point. In 1997, Silver considers Torrance's (1974) categories of *fluency*, *flexibility*, and *originality* for the assessment of problem-posing products, which are the posed problems. In the context of problem posing, fluency refers to the number of posed problems, flexibility refers to the diversity of posed problems (e.g., in terms of different mathematical ideas or strategies to be applied), and originality refers to the rareness of the posed problems compared within the solution space of a peer group .

A review on the connection between problem posing and mathematical creativity can be found in Joklitschke et al. (2019). They found only eleven articles in all databases of high-ranked journals on mathematics education research investigating the connection between problem posing and mathematical creativity. Furthermore, they identified three clusters within these articles.

1. Articles in which problem-posing situations are used to foster creativity (Haylock, 1997; Bonotto, 2013; Sriraman & Dickman, 2017). Sriraman and Dickman (2017), for example, use *mathematical pathologies*, the Lakatosian heuristic, and problem-posing activities to address students' creativity.

2. Articles in which problem posing is used to identify and investigate creativity via Guilford's (1967) and Torrance's (1974) framework (Bonotto, 2013; Van Harpen & Presmeg, 2013; Van Harpen & Sriraman, 2013; Leung, 1997). In general, these studies find that problem posing is suitable for investigating the creative abilities of participants. On the other hand, they find a correlation between problem posing and creative abilities.

3. Articles in which other approaches are used to measure creativity through problem posing (Voica & Singer, 2013; Singer et al., 2017). These studies use organizational theory to investigate the relationship between cognitive flexibility and problem-posing abilities. The participants' abilities on how the posed problems are coherent and consistent serves as a measure for this relationship.

Renzulli (2005; Renzulli & Reis, 2021) famously conceptualizes creativity as a part of his seminal Three-Ring model for giftedness (see Figure 5.8 on the following page). He describes that giftedness is based on three interacting clusters of traits, namely *creativity*, *above-average ability*, and *task commitment*.

There are only few empirical studies that investigate the relationship between problem posing and giftedness (Arıkan & Ünal, 2015; Pelczer,

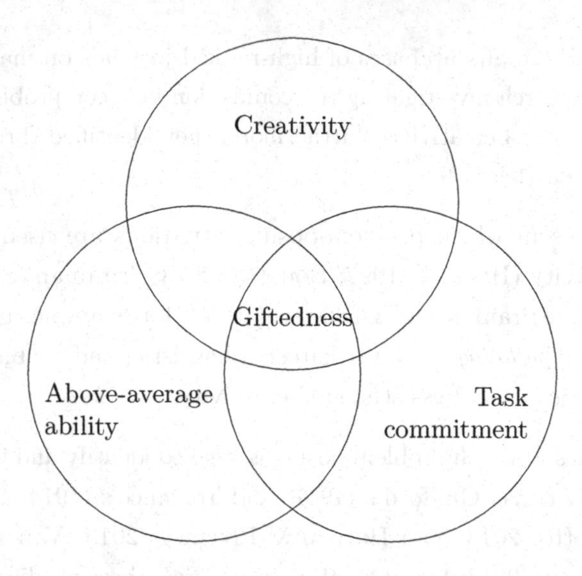

Fig. 5.8.: Three-Ring model of Giftedness by Renzulli (2005)

2008; Runco & Nemiro, 1994; Erdoğan & Gül, 2020). However, the approaches with regard to the chosen research methods and problem-posing activities of these studies are too different to make a general tendency regarding the relationship between problem posing and giftedness. It can be concluded that research in this area may be necessary.

Part III

Journal articles

Rethinking problem-posing situations: A review

6

The Version of Record of this manuscript has been published and is available on https://doi.org/10.1080/19477503.2020.1841501. Full reference: Baumanns, L., & Rott, B. (2021). Rethinking problem-posing situations: a review. *Investigations in Mathematics Learning*, *13*(2), 59–76.

L. Baumanns, *Mathematical Problem Posing*, Kölner Beiträge zur Didaktik der Mathematik, https://doi.org/10.1007/978-3-658-39917-7_6

Abstract: *In research on mathematical problem posing, a broad spectrum of different situations is used to induce the activity of posing problems. This review aims at characterizing these so-called problem-posing situations by conducting three consecutive analyses: (1) By analyzing the openness of potential problem-posing situations, the concept of ‚mathematical posing' is concretized. (2) The problem-posing situations are assigned to the categories free, semi-structured, and structured by Stoyanova and Ellerton to illustrate the distribution of situations used in research. (3) Finally, the initial problems of the structured problem-posing situations are analyzed with regard to whether they are routine or non-routine problems. These analyses are conducted on 271 potential problem-posing situations from 241 systematically gathered articles on problem posing. The purpose of this review is to provide a framework for the identification of differences between problem-posing situations.*

6.1 Introduction

In 1980, the mathematician Paul Halmos published an article about his idea of "the heart of mathematics". In the epilogue, he wrote: "I do believe that problems are the heart of mathematics, and I hope that as teachers, in the classroom [...] we will train our students to be better problem-posers and problem-solvers than we are" (1980, p. 524). This way, he highlighted two activities: problem solving and problem posing. Both have been emphasized as important by mathematicians (Pólya, 1957; Cantor, 1966; Lang, 1989; Tao, 2006) and mathematics educators (Brown & Walter, 1983; Schoenfeld, 1985b; Silver, 1994; English, 1997). However, whereas the problem solving received a lot of attention in the last decades of mathematics education research, especially since Pólya's (1957) and Schoenfeld's (1985b) seminal works, problem posing "has not been a major focus in mainstream mathematics education research" (Singer et al., 2013, p. 4). Problem posing is an important companion of problem solving, it can encourage flexible thinking, improve problem-solving skills, and sharpen learners' understanding of mathematical contents (English, 1997). In school contexts, especially Brown's and Walter's (1983) approach forms a foundation for many teachers to initiate rich problem-posing activities. Consequently, problem posing should be given a relevant status in research of mathematics education.

Due to a lack of research on theoretical foundations, "the field of problem posing is still very diverse and lacks definition and structure" (Singer et al., 2013, p. 4). In order to structure this field, it, therefore, seems advisable to review empirical studies and theoretical contributions dealing with this topic. To underline one aspect in which we think the field lacks definition and structure, we start our discussion with two inducements to pose problems (see Table 6.1). In this article, these inducements are called *potential problem-posing situations*; please note that the supplement potential is added because all situations are examined in the extent to which they induce problem posing. This article focuses on these kind of analyses of potential problem-posing situations.

The situations in Table 6.1 were used in studies that investigate the relationship between problem solving and problem posing (Cai & Hwang, 2002; Arıkan & Ünal, 2015). Cai and Hwang (2002) found a strong link between problem solving (as the ability to successfully solve non-routine problems, the participants' use of strategies, and their solution's representation) and problem posing (as the ability to pose a variety of problems that extend the initial problem). Similarly, Arıkan and Ünal (2015) found a strong correlation ($r = 0.76$; $p < 0.01$) between problem solving (as the ability in finding and producing multiple solutions) and problem posing (as the ability to answer situations similar to Situation 2 in Table 6.1). However, a comparison of the potential problem-posing situations reveals differences which we want to outline in the following.

	Situation
1	Mr. Miller drew the following figures in a pattern, as shown below. 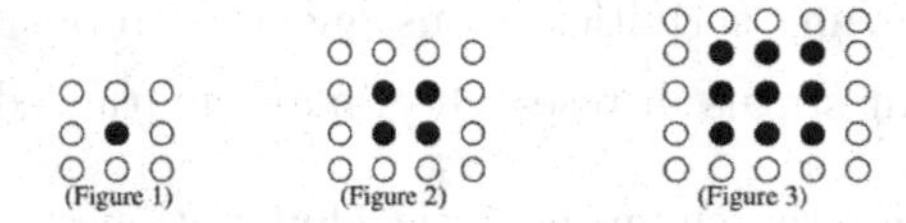 For his student's homework, he wanted to make up three problems based on the above situation: an easy problem, a moderate problem, and a difficult problem. These problems can be solved using the information in the situation. Help Mr. Miller make up three problems and write these problems in the space below. (Cai & Hwang, 2002, p. 405)
2	Which one of the below problems can be matched with the operation of $213 + 167 = 380$? A) Osman picked up 213 pieces of walnut. Recep picked up 167 more pieces of nuts more [sic] than Osman. What is the total amount of the nuts that both Osman and Recep picked up? B) On Saturday, 213 and on Sunday 167 bottles of water were sold in a market. What is the total number of bottles of water that were sold at this market on these two days? C) Erdem has 213 Turkish lira. His brother has 167 lira less than that. What is the total amount of money that both Erdem and his brother have? (Arıkan & Ünal, 2015, p. 1410)

Tab. 6.1.: Potential problem-posing situations used to investigate the relationship between problem posing and problem solving.

Cai and Hwang (2002; Situation 1 in Table 6.1) invite students to pose at least three problems with varying degrees of difficulty for given dot patterns. This can result in numerous different problems, some of which will likely be non-routine problems, such as the following ones given by Cai and Hwang: "How many white dots are there in the twentieth figure?", "How many more black dots are there in the tenth figure than in the fifth figure?", or "How many black dots are there in the first five figures?" (2002, p. 413). Situation 2 from Arıkan and Ünal (2015; see Table 6.1) is taken from a mathematics textbook for third-graders and served as an infrastructure for their study. The situation provides a defined calculation and three problems of which only answer B) applies to the given calculation. Therefore, there is only one correct answer and the situation does not encourage posing further problems.

Thus, although both studies come to a similar conclusion, we argue that they do not refer to the same relationship because the situations' characteristics may initiate divergent problem-posing activities. Such differences between potential problem-posing situations motivated a more in-depth investigation. Therefore, we conducted a document analysis on potential problem-posing situations that are used in empirical studies or as illustrative examples in theoretical contributions. The aim of this article is to offer a new perspective on these potential problem-posing situations that considers differences like the ones shown in Table 6.1. This perspective is developed by three consecutive analyses that classify all potential problem-posing situations. The following section describes the theoretical constructs needed for these three analyses.

6.2 Theoretical Background

To characterize potential problem-posing situations, current understandings of problem posing are presented. Afterwards, problem posing is broken down into its etymological components, *posing* and *problem*, which are defined and discussed individually.

6.2.1 What is problem posing?

There are two prominent definitions of the term *problem posing*, at least one of which is used or referred to in the majority of the reviewed research articles on the topic. The first one was proposed by Silver (1994, p. 19), who describes problem posing as the generation of new problems and reformulation of given problems which can occur *before*, *during*, or *after* a problem-solving process. Posing *before* problem solving refers to the generation of a new problem that is to be solved afterwards. Posing a problem *after* problem solving refers to Pólya's (1945) phase of "Looking Back". At last, posing a similar and simpler problem *during* problem solving may help to find a solution as suggested in Pólya's "Devise a Plan"- or Schoenfeld's (1985b) "Exploration"-phase. In each of these cases, posing activities are intended at developing problems that demand a solution. The second definition comes from Stoyanova and Ellerton (1996, p. 518), who refer to problem posing as the "process by which, on the basis of mathematical experience, students construct personal interpretations of concrete situations and formulate them as meaningful mathematical problems".

Both definitions of problem posing can be applied to a wide spectrum of situations. In this article, we adopt the definition of Stoyanova and Ellerton (1996) as our underlying understanding, because it particularly emphasizes the interplay between the problem poser and the problem-posing situations. However, both definitions are not disjunctive, but describe equivalent activities. Following Stoyanova (1997, p. 5), a situation is a not well-structured problem in the sense that the goal cannot be determined by all given elements and relationships. From this perspective, a problem-posing situation is a situation that includes the activity of problem posing.

6.2.2 What is posing?

In the research literature on problem posing, the term posing has not been examined in much detail. Therefore, we consulted dictionaries. Cambridge Dictionary (2020) defines "to pose" as "to ask a question". We will refer to this understanding of posing as some of the situations analyzed hereafter do not induce posing in this sense, i.e. there is no question to be asked or the raised question is already solved (cf. Situation 2 in Table 6.1).

6.2.3 What is a problem?

In mathematics education research, the term *problem* has been (and still is being) used in a variety of meanings, leading to some difficulties and misinterpretations in reading the research literature Schoenfeld (1992, p. 337). While recent studies on *mathematical problem solving* consistently investigate non-routine problems, the term *problem* in *problem posing* still refers to any kind of mathematical task. Therefore, we first conceptualise mathematical tasks in general, and then highlight a perspective that will be used in this article to analyse problem-posing situations.

A mathematical task is any kind of artifact that aims at focusing students' attention on a specific mathematical idea (Stein et al., 1996). Different attempts to conceptualise tasks (Newell & Simon, 1972; Wickelgren, 1974; Doyle, 1988) consistently identify three components that tasks consist of: given expressions (initial state), goal expressions and desired outcome (goal state), and operations that transform the initial state into the goal state (operators). Depending on the given information of a task regarding these three components, different types of tasks can be characterised. Table 6.2 summarizes different types of tasks.

Typs of tasks	Initial state	Operators	Goal state	Examples of sources that contain theoretical conceptualisations and/or empirical studies in the particular type of task
(1) Worked task	×	×	×	Sweller and Cooper (1985); Van Gog et al. (2006)
(2) Routine tasks	×	×	–	Díez and Moriyón (2004)
(3) Non-routine tasks	×	–	×/–	Multiple solution tasks (Leikin & Lev, 2013) or comparing use of strategies (Posamentier & Krulik, 2008; Koichu et al., 2007)
(4) Reversed tasks	×/–	–	×	Groetsch (2001)
(5) Open tasks	×/–	–	–	Open-ended problems (Pehkonen, 1995; Cifarelli & Cai, 2005), exploration (Mason et al., 2010), or problem posing (Silver, 1994; Cai & Hwang, 2002)

Tab. 6.2.: Types of tasks with regard to the known (×) or unknown (–) information concerning the initial state, the operators, and the goal state (Bruder, 2000; Leuders, 2015)

For some of these types of tasks, like the (1) worked tasks or (2) routine tasks, an individual may have “ready access to a solution schema” (Schoenfeld, 1985b, p. 74), i.e. knows a procedure for transforming the initial into the goal state. Solving these is a routine activity. For others, like the (3) non-routine tasks or (5) open tasks, an individual may have no access to a solution schema. Solving these tasks is a non-routine activity that involves understanding the initial and goal state by interpreting the problem’s situation mathematically (Lesh & Zawojekski,

2007). Solving tasks is here conceptualised as an activity with varying degrees of routineness. The term *problem* in *problem posing* also refers to all types of tasks with differing degrees of routineness (Cai & Hwang, 2020; Baumanns & Rott, 2019). This understanding is used in this article for the analysis and, therefore, should be given concrete terms. We terminologically follow Pólya (1966) and specify the term *problem* with the supplements *routine* and *non-routine* within this article.

Let us exemplify this difference: Integrating a polynomial function of degree 53 can be a quite tedious computational activity for mathematicians; however, since they know methods of integrating polynomials, this is a routine problem for them. Otherwise, the problem to determine the number of squares on a regular chessboard might be a non-routine problem to some mathematicians, if the method of how to calculate this number is currently unknown to them.

In most cases, the decision of whether a given problem is a routine or non-routine problem is evident. Nevertheless, this attribution is specific to individuals; a problem that is non-routine for one person can be routine for another person who knows an apt solution scheme (Schoenfeld, 1989; Rott, 2014). Thus, the demarcation between these categories may not be sharp but the extreme cases are clearly recognizable (Pólya, 1966, p. 126).

With regard to the question whether a problem-posing activity is itself a problem-solving activity, different positions are taken within the literature. While some researchers identify problem posing as a special case of problem solving (Silver, 1995; Kontorovich et al., 2012; Arıkan & Ünal, 2015), we, along with others, identify both as similar but characteristically different activities (Pelczer & Gamboa, 2009; Dickman, 2014). However, we want to emphasize that the attributions of the categories routine and non-routine problem relate exclusively to initial or posed problems and not to the activity of the problem posing itself.

6.2.4 Openness of problems

The above examples of routine and non-routine problems are closed in the sense that they only have one correct answer (Becker & Shimada, 1997). In comparison, problem-posing situations are mostly considered as open (Pehkonen, 1995; Silver, 1995; Cifarelli & Cai, 2005) because there are several possible answers to a problem-posing situation. To assess the openness of the potential problem-posing situations, we refer to the five task variables provided by the framework of Yeo (2017): *goal*, *method*, *task complexity*, *answer*, and *extension*. Each of these is discussed with regard to whether it is *closed* or *open*. Yeo (2017) suggests to use this framework to assess the openness of problem-posing situations. The variables *answer* and *extension* are particularly related to problem posing and for the examined potential problem-posing situations, noteworthy differences have been identified with regard to these variables. Therefore, we outline and exemplify them in the following to use them in the further analyses:

Answer: A problem has a closed answer if it is possible to determine all correct answers, e.g. for the problem "Solve the quadratic equation $x^2 + 2x - 3 = 0$" (Yeo, 2017, p. 177) all solutions can be determined. A problem has an open answer, if it is not possible to determine all correct answers. Open answers are well-defined if it can objectively be decided if they are correct or incorrect, e.g. for the problem "Power of 3 are $3^1, 3^2, 3^3, 3^4, \ldots$ Investigate" (Yeo, 2017, p. 178) it can be stated that the investigations are correct with regard to the powers of 3. Open answers are ill-defined if the answer to a problem can be determined as valid on a subjective level; e.g. the problem "Design a playground for the school" (Yeo, 2017, p. 117) does not really have a correct or incorrect answer.

Extension: If a problem cannot or should not be extended, e.g. because extending a problem leads to new unrelated problems, it is considered to be closed. For example, the problem of designing a swimming pool is not an extension of the problem of designing a playground, but a completely new problem (Yeo, 2017, p. 186). If the extension leads to new related problems that help to discover more underlying patterns or mathematical structures, it is considered to be open. This openness can either be subject-dependent if the problem itself does not explicitly have an instruction to raise new problems and it depends on the subject whether new problems are raised, e.g. for the problem to solve the quadratic equation stated before. The openness can also be task-inherent if the student is expected to extend the given problem, e.g. when investigating the powers of 3 as in the problem above, students are expected to look for further patterns related to the powers of 3 or even for other numbers.

6.2.5 Categorisations of problem-posing situations

As one of the most established categorizations, Stoyanova and Ellerton (1996) differentiate between *free*, *semi-structured*, and *structured* problem-posing situations, depending on their degree of given information. Free situations provoke the activity of posing problems out of a given, naturalistic or constructed situation without any restrictions (see Table 6.3, Situation 3). In semi-structured situations, posers are invited to explore the structure of an open situation by using mathematical knowledge, skills, and concepts of previous mathematical experiences. Stoyanova (1997; 1999) further differentiates between semi-structured situations which are based on a specific problem structure (see Table 6.3, Situation 4) and semi-structured situations which are based on a specific solution structure (see Table 6.3, Situation 5). In structured situations, people are asked to pose further problems based on a specific problem, e.g. by varying its conditions. Structured situations are further differentiated between situations which are based on a specific problem (see Table 6.3, Situation 6) and situations which are based on a specific solution (see Table 6.3, Situation 7).

Based on the framework by Stoyanova and Ellerton (1996), Christou, Mousoulides, Pittalis, Pitta-Pantazi, and Sriraman (2005, p. 151) postulate five categories of problem-posing situations: to pose (a) a problem in general, (b) a problem with a given answer, (c) a problem that contains certain information, (d) questions for a problem situation, and (e) a problem that fits a given calculation.

Cai and Jiang (2017) used the classification by Christou, Mousoulides, Pittalis, Pitta-Pantazi, and Sriraman (2005) to identify different types of problem-posing situations in five textbook series. They came up with four categories depending on what the problem requires students to do, in relation to the information contained in the problem: (1) Posing a problem that matches the given arithmetic operation(s), (2) posing variations of a question with similar mathematical relationship or structure, (3) posing additional questions based on the given information and a sample question, and (4) posing questions based on given information.

However, the characteristics of these categories differ: The activities induced by the rather open situations in Table 6.3 aim at investigating patterns and structures of an underlying mathematical content by exploring multiple mathematical pathways. Compared to that, investigating is not induced by some representatives of the categories by Christou, Mousoulides, Pittalis, Pitta-Pantazi, and Sriraman (2005) and Cai and Jiang (2017). For example, "Make up a word problem orally for $14 + 8 = ?$" (Cai & Jiang, 2017, p. 1526) is a representative of the category *Posing a problem that matches the given arithmetic operation(s)*. We argue that reacting to this in a way like "Anne has 14 and Ben 8 apples. How many do they have in total?" is only providing a context, but not posing a problem and should, therefore be distinguished from the investigations induced by the situations in Table

		Category	Situation
3		**Free**	"Give an example of a problem which can be solved by finding the Least Common Multiple of two or more integers" (Stoyanova, 1997, p. 64)
4	**Semi-structured**	**Based on a specific problem structure**	"Consider the following infinite sequence of digits: 1234567891011121314 15... 999100010011002... Note that it is made by writing the base ten counting numbers in order. Ask some meaningful questions. Put them in a suitable order." (Stoyanova, 1999, p. 32)
5	**Semi-structured**	**Based on a specific solution structure**	"Finish the problem situations below so that the solution method implies the use of permutation: Two girls and four boys are standing in a line." (Stoyanova, 1997, p. 68)
6	**Structured**	**Based on a specific problem**	"Consider the sequence $1, 2, 3, 4, 5, \ldots N$. If $N = 200$, how many digits have been used? Other questions?" (Stoyanova, 1999, p. 34)
7	**Structured**	**Based on a specific solution**	"Find the sum of: $1 - 2 + 3 - 4 + 5 - \ldots + 999 - 1000$. [Suggest similar problems and predict possible links between this similarity and an expected solution approach.]" (Stoyanova, 1999, p. 35)

Tab. 6.3.: Free, semi-structured, and structured problem-posing situations

6.3. These noteworthy differences between these situations, which in the literature are all referred to with the same term, indicate the need for an in-depth analysis of potential problem-posing situations, which leads to our research objective.

6.2.6 Research questions

Our aim is to propose a framework for the identification of differences between situations. The following consecutive questions serve developing this framework:

(1) What potential problem-posing situations induce the activity of posing? How open are the potential problem-posing situations?

To answer these questions, we will refer to the presented understanding of posing as well as Yeo's (2017) framework regarding the openness of problems?

(2) Referring to the posing situations resulting from question (1), how well does the further categorization of those situations into *free*, *semi-structured* and *structured* situations succeed?

Responding to this question, we will analyze the situations' distribution by trying to apply the framework by Stoyanova and Ellerton (1996), distinguishing situations into *free*, *semi-structured*, and structured.

(3) Referring to the structured situations resulting from question (2), to what extent can the initial problems of those situations be assigned to the categories *routine* and *non-routine problem*?

This question is answered by analyzing the initial problems of structured problem-posing situations with regard to whether they are routine or non-routine problems. It is further discussed what influence the initial problem may have on the posed problems.

6.3 Design

To answer the research questions, we have systematically gathered articles, adopting the guidelines for a systematic literature review by Moher et al. (2009). With the situations of these articles, a document analysis was conducted (Bowen, 2009) that aims at analyzing and comparing potential problem-posing situations with the goal to identify their characteristics which could be of interest to mathematics educators and researchers on mathematics education.

6.3.1 Data sampling

We gathered articles on the topic of problem posing from four different sources: (a) Top-ranked journals in mathematics education research, (b) the Web of Science, (c) proceedings of the Conferences of the International Group for the Psychology of Mathematics Education (PME), and (d) edited books on problem posing from the international mathematics education community. We gathered both empirical studies as well as theoretical articles.

(a) We gathered articles from all eleven A*-, A-, and B-ranked journals in mathematics education (as classified by Törner and Arzarello (2012, p. 53)).[1] In each of the journal's databases, we searched in all available years from its initiation up to 2019 with the search

[1] Name of the journals and their founding year: Educational Studies in Mathematics (from 1969), Journal for Research in Mathematics Education (from 1970), For the Learning of Mathematics (from 1980), Journal of Mathematical Behavior (from 1980), Journal of Mathematics Teacher Education (from 1998), Mathematical Thinking and Learning (from 1999), ZDM Mathematics Education (from 1997), International Journal of Mathematical Education in Science and Technology (from 1970), International Journal of Science and Mathematics Education (from 2003), Mathematics Education Research Journal (from 1989), and Research in Mathematics Education (from 1999). These journals are identical to the eleven top-ranked journals in mathematics education research from the opinion-based study by Williams and Leatham (2017, p. 390).

term "problem posing" which led to 686 articles. After that, only those articles were extracted that have the term "problem posing" in either their title, abstract, or keywords which lead to 74 articles from these top-ranked journals.

(b) In the database *Web of Science*, we used the search term "problem posing" for all available years from 1945 (which is as far back as the database reaches) to 2019 in the title, abstract, or keywords within selected categories on mathematics and its education. Excluding the already considered articles from the A*-, A-, and B-ranked journals, this led to another 91 articles.

(c) Moreover, we searched the proceedings of the Conferences of the International Group for the Psychology of Mathematics Education (PME) from the years 2000 to 2018 using the search term "problem posing" in the title. The criterion of having the term "problem posing" in the abstract or keywords of the article was omitted because not every PME article offers an abstract and none offer keywords. This led to 44 additional articles.

(d) Finally, we gathered all 26 articles of the book edited by Singer et al. (2015) as well as 7 articles of the book edited by Felmer et al. (2016) that have the term "problem posing" either in their title, abstract, or keywords.

In total, 241 articles on problem posing have been reviewed which contain 271 situations (see Figure 6.1 on the next page). We are aware that some of these articles deal with different aspects of the same studies and, thus, some situations may occur multiple times. However, possible doublings do not affect the central statements of the following analyses.

The potential problem-posing situations that are presented in the analyses were not chosen because of their quality; we neither argue for nor against them being suitable for research or practice. However, the

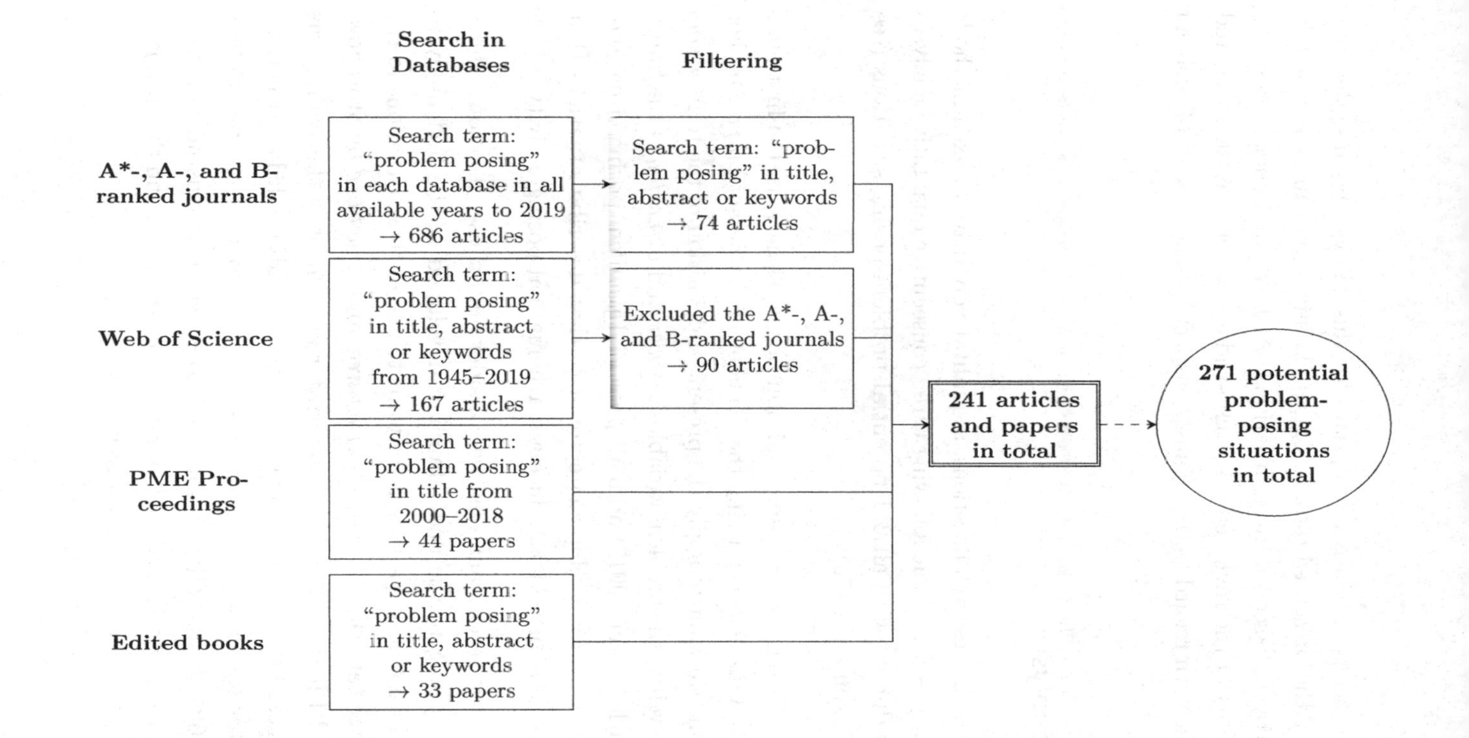

Search in Databases
Filtering
A*-, A-, and B-ranked journals
Search term: "problem posing" in each database in all available years to 2019 → 686 articles
Search term: "problem posing" in title, abstract or keywords → 74 articles
Web of Science
Search term: "problem posing" in title, abstract or keywords from 1945–2019 → 167 articles
Excluded the A*-, A-, and B-ranked journals → 90 articles
PME Proceedings
Search term: "problem posing" in title from 2000–2018 → 44 papers
Edited books
Search term: "problem posing" in title, abstract or keywords → 33 papers
241 articles and papers in total
271 potential problem-posing situations in total

Fig. 6.1.: Procedure for the selection of articles

situations are particularly appropriate for this analysis because the arguments for the presented terminological classification can be adequately demonstrated based on these situations. At last, we do not discuss the conclusions of the reviewed research studies, which use the presented situations with regard to the cognitive aspects of the process of problem posing.

6.3.2 Data analysis

To pursue the research questions, a qualitatively driven mixed-methods approach was used. The following three consecutive qualitative analyses were conducted with all 271 potential problem-posing situations (see Figure 6.2 on page 106):

(1) By distinguishing posing and non-posing situations, the first analysis aims at concretizing the concept of *mathematical posing*. For that, we want to apply the presented definition of the term *posing* as well as the problem variables *answer* and *extension* of the framework by Yeo (2017) onto all potential problem-posing situations. We want to develop characteristics that describe situations that either do or do not induce a mathematical posing activity.

(2) For the second analysis, we consider only those situations which, according to the understanding developed in the first analysis, induce a posing activity. We want to assign these situations to the three categories *free*, *semi-structured*, and *structured* by Stoyanova and Ellerton (1996). The interrater agreement of this coding was checked through a second rater. This rater also coded all situations after a one-hour training. Calculating Cohen's Kappa (Cohen, 1960) of this coding is intended to check the robustness of these three categories and to illustrate the distribution and the spectrum of situations used in research.

(3) In a third analysis, we look exclusively at those situations that were categorized as structured by both raters in the second analysis. The initial problems of these situations are examined with regard to Pólya's (1966) distinction between *routine* and *non-routine* problems. Free and semi-structured situations are not considered for this analysis as they do not include initial problems that can be examined. So far, research on problem posing has hardly considered the initial problems of structured situations, although it is assumed that the initial problem's characteristics could influence the problems posed.

After coding the situations using these analyses, the codes were evaluated quantitatively in order to illustrate the distribution of the developed categories.

6.4 Analysis 1: Is it posing?

For Analysis 1, all 271 potential problem-posing situations have been examined with regard to whether they induce a posing activity at all. Furthermore, from Yeo's (2017) framework to characterize the openness of mathematical problems, the variables *answer* and *extension* are used to examine the openness of the potential problem-posing situations. This analysis aims at concretizing the concept of *mathematical posing*. This concretization is developed using the situations shown in Table 6.4 on page 107.

Situation 7 invites providing a context to a given calculation. The pictures show either three bags with four candies or four bags with three candies in each of them. For both pictures, the students are asked to generate a partitive and a quotitive division scenario. For the first figure, such a scenario could be: "Twelve candies are separated into three bags so that each bag has four candies."

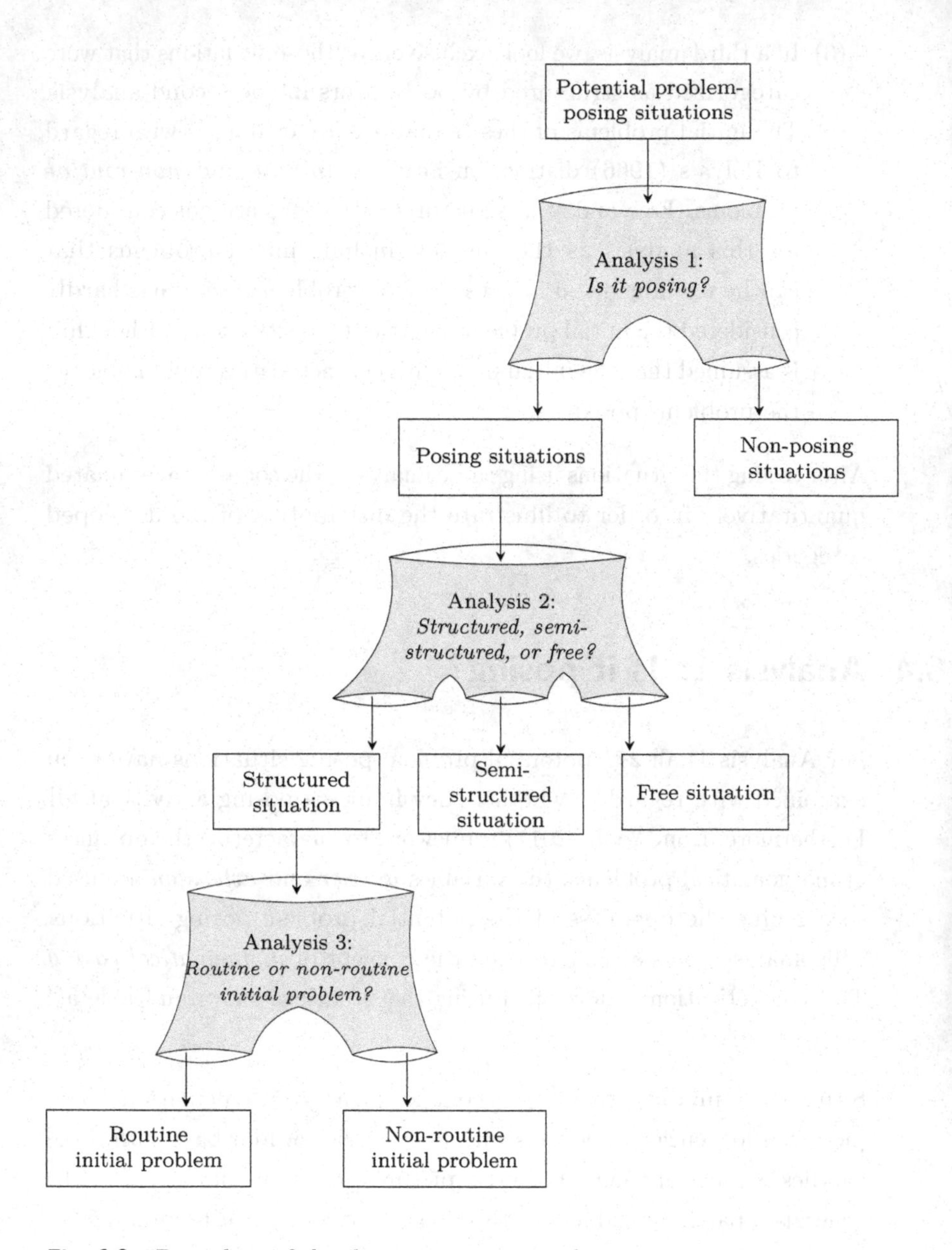

Fig. 6.2.: Procedure of the three consecutive analyses

In situation 8 by Silber and Cai (2017), students are invited to search for the question to a given context and its answer. The sought-after problem, however, is predefined: "How many miles did Jerome, Elliot, and Arturo drive altogether?"

	Situation
7	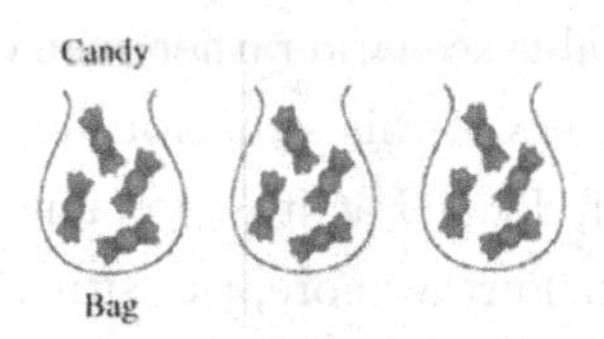 Generate one division scenario consistent with the given picture and the numerical equation 12/3 = 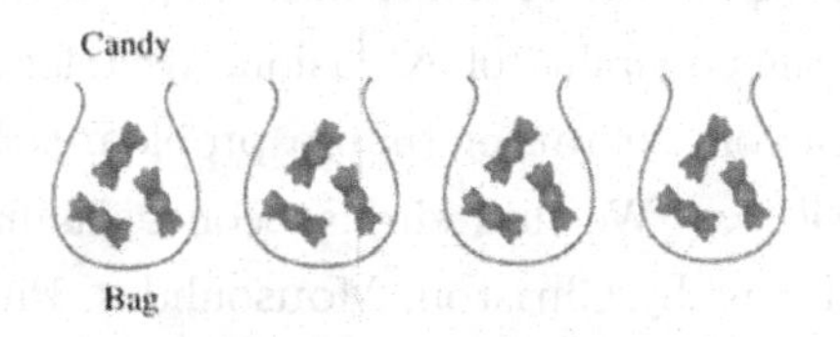 Generate one division scenario consistent with the given picture and the numerical equation 12/3 = (Kinda, 2013, p. 193)
8	Jerome, Elliot, and Arturo took turns driving home from a trip. Arturo drove 120 miles less than Elliot. Elliot drove twelve times as many miles as Jerome. Jerome drove 50 miles. Write a problem that has "1130 miles" as the answer. (Silber & Cai, 2017, p. 168)

Tab. 6.4.: Potential problem-posing situations discussed regarding the term posing.

6.4.1 Discussion of Analysis 1: Mathematical problem posing

When providing a context that matches a given calculation or given data, as in situation 7, no question arises which can be solved afterwards. And even if someone poses a question to the given calculation, such as "If you put 12 candies in 3 bags, how many candies are in each bag?" for which the answer is already given in the figure, the focus of this problem is on creating a suitable scenario rather than on posing a question. With regard to the openness of this situation (cf. Yeo, 2017), the answer to this situation is well-defined as it is possible to determine whether it is correct or incorrect. Furthermore, the situation cannot be extended to new related problems, e.g. by changing the numerical equation as well as the picture. This does not lead to new problems that help to discover more underlying patterns or mathematical structures, thus, it is closed with regard to the possibility of extension. We refer to situations like this one, in which a context rather than a problem is asked for, as *context providing problems.* We find similar conceptualizations like context providing problems by Christou, Mousoulides, Pittalis, Pitta-Pantazi, and Sriraman (2005; *posing a problem that fits a given calculation*) or Cai and Jiang (2017; *posing a problem that matches the given arithmetic operation(s)*).

Reacting to situation 8 in an expected way leads to posing a problem. However, since the situation offers a result and has an undefined question, it can rather be referred to as a *reversed problem.* Such problems are often used in school contexts as an educational tool (Bruder, 2000), yet solving a reversed problem is not a posing activity. Instead, it is the activity of reconstructing an implicitly existing question based on a given solution. In addition, solving a reversed problem does not lead to a problem that demands a solution because it has already been solved. Thus, the answer to a reversed problem is closed because it has a particular solution (Silber & Cai, 2017). Furthermore, reversed problems do not have an adequate possibility of extension that leads to new problems and help to discover underlying patterns or mathematical structures. Thus, they are also closed with regard to the possibility of expansion. Nevertheless, the relevance of solving reversed problems as a didactical instrument should not be denied. Again, we find similar conceptualizations like reversed problems by Christou, Mousoulides, Pittalis, Pitta-Pantazi, and Sriraman (2005; *posing a problem with a given answer*) or Cai and Jiang (2017; *posing variations of a question with similar mathematical relationship or structure*).

Based on these analyses and in addition to the presented understanding of *posing*, we, therefore, specify the meaning of *mathematical posing*. As mentioned before, Silver (1994, p. 19) states that posing activities can occur *before*, *during*, or *after* problem solving. In each of these cases, posing activities are intended at developing problems that demand a solution. Therefore, we add to the concept of *mathematical posing* that this activity aims at developing problems that demand a solution.

Both context providing problems and reversed problems have in common that they have a solution for which it can be decided whether it is correct or incorrect on an objective level. Thus, they are at least well-defined with regard to the openness of the answer. In problem posing as a process of interpreting concrete situations (Stoyanova & Ellerton, 1996, p. 518), it is not expedient to refer to a posed question as correct or incorrect. It may be reasonable to emphasize whether the posed problems are valid on a subjective level which Yeo (2017) refers to as ill-defined answers.

Furthermore, context providing problems and reversed problems are closed with regard to the possibility of extension. However, the purpose of an extension is often to discover underlying patterns, mathematical structures, or possible generalizations. Therefore, the possibility of extension is added to the concept of mathematical posing. Most problem-posing situations have a task-inherent option of extension (e.g. Table 6.3).

Whether a given setting is either a non-posing or a posing situation can be determined a priori by analyzing whether the particular characteristics presented above apply. In total, of the 271 analyzed situations, 79 are context providing problems and 16 are reversed problems. Hence, with the presented understanding of mathematical posing, 35.1 % of the analyzed situations are not considered as mathematical posing situations. We argue that this distinction between posing- and non-posing activities contributes to the field of problem posing because the analyses illustrate that both activities differ characteristically from each other. As Ruthven (2020) said, asking students to solve modified problems like reverse versions of standard problems may carry the risk of a superficial interpretation of problem posing. Therefore, researchers investigating the activity of problem posing should bear in mind what activity they want to investigate. This should also be considered when reviewing research results.

6.5 Analysis 2: Is it free, semi-structured, or structured?

After identifying situations that do not induce posing activities in Analysis 1, 176 situations remain that do induce mathematical posing activities. These situations are now categorized using the framework by Stoyanova and Ellerton (1996; see Table 6.3). Some of the situations had been assigned to these categories within the articles they originated. However, as not every situation had been categorized and not every categorization had been explained, we (hereafter referred to as Rater 1) coded all situations, ignoring previous codings if they existed. For assessing the interrater-agreement, an additional and independent rater (hereafter referred to as Rater 2) also categorized all situations. Rater 2 is an undergraduate student who deals with problem posing as part of a thesis. Rater 2 was introduced to the coding by the authors in a one-hour training session in which anchor examples of the individual analyses were considered. After about 10 % of the situations had been coded, a second training session was held to discuss and eliminate the doubtful cases before the rest of the situations were coded. In our analyses, the coding of Rater 1 is taken as the valid coding. This analysis aims at providing information about the distribution of the named categories of problem-posing situations in research.

Table 6.5 shows the agreement table for Analysis 2. In total, 32 of the 176 posing situations (~18.1 %) were based on an initial problem and were thus coded as structured situations, 96 (~54.5 %) situations were coded as semi-structured situations, and 48 (~27.3 %) as free situations by Rater 1. The interrater-agreement is excellent (Fleiss et al., 2003) with a Cohen's Kappa of $\kappa = 0.79$. However, there are noteworthy differences between the agreement of the individual categories. Of the 32 situations that Rater 1 coded as structured, Rater 2 deviated in only one case (~3.1 %), for semi-structured situations, Rater 2 deviates in 7 of 96 situations (~7.3 %), and for free situations even in 22 of 48 situations (~45.8 %). The latter deviations mainly concern the separation of free and semi-structured problem-posing situations. As an example, the following situation by Tichá and Hošpesová (2013) was coded as free by Rater 1 and as semi-structured by Rater 2: "Pose three word problems which include fractions 1/2 and 3/4. Solve the posed problems."

6.5.1 Discussion of Analysis 2: Unstructured situations

Table 6.5 shows that, in contrast to structured situations, the coding of free and semi-structured situations has a lower agreement. In the consensual discussion of coding between the two raters, it was noticeable

Rater 2	Rater 1: Free	Semi-structured	Structured	Non-posing	Total	p
Free	26	5	0	0	31	0.11
Semi-structured	20	89	0	10	119	0.44
Structured	0	0	31	0	31	0.11
Non-posing	2	2	1	85	90	0.33
Total	48	96	32	95	271	1.0
p	0.18	0.35	0.12	0.35	1.0	**0.79**

Tab. 6.5.: Agreement table for coding all situations into free, semi-structured, structured, and non-posing situations.

that the descriptions by Stoyanova and Ellerton (1996) did not allow a clear attribution of the differently coded situations. In fact, even the originators of this categorization were inconsistent in the assignment: Stoyanova and Ellerton (1996, p. 523) exemplify semi-structured situations by a problem to ask students to pose problems related to a right-angled triangle. In a later publication, Stoyanova (1997, p. 64) categorizes the same situation as free.

As an example from the consensual discussion to strengthen our argument that semi-structured and free situations are hard to tell apart, consider the following problem-posing situation by Cai et al. (2015, p. 7): "Ann has 34 marbles, Billy has 27 marbles, and Chris has 23 marbles. Write and solve as many problems as you can that use this information." On the one hand, because of the limited information provided, it can be coded as a free situation. On the other hand, the number of marbles must be analyzed number-theoretically and within the given context in order to pose meaningful problems; thus, it can also be justified that this is a semi-structured situation.

We, therefore, propose that free and semi-structured situations should not be understood as clearly distinguishable categories. Instead, representatives of these categories should be placed on a spectrum of *unstructured* situations. Unstructured situations refer to a spectrum of situations with less restrictions (see Table 6.6, Situation 9) to situations with extensive information (see Table 6.6, Situation 11). Structured

		Situation
9	Unstructured	[W]rite (or tell me about) a mathematics problem you would like to solve. (Lowrie, 2002, p. 91)
10	Unstructured	[Pose a problem to the following situation:] A piece of cake is cut into 8 equal parts. Wah ate 4/8 piece[s], Ming ate 2/8 piece[s], Kong ate 2/8 piece[s]. (Leung, 2013, p. 108)
11		Imagine billiard ball Tables like the ones shown below. Suppose a ball is shot at a 45° angle from the lower left corner (A) of the Table. When the ball hits a side of the Table, it bounces off at a 45° angle. In Table 1, the ball travels on a 4×6 Table and ends up in pocket B, after 3 hits on the sides. In Table 2, the ball travels on a 2×4 Table and ends up in pocket B, after 1 hit on the side. In each of the figures shown below, the ball hits the sides several times and then eventually lands in a corner pocket. 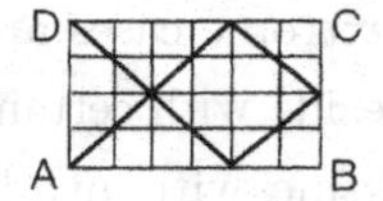 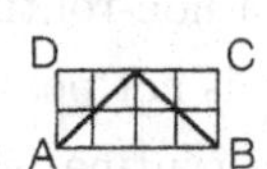 Based on the given situation, pose as many interesting mathematical problems as you can. (Koichu & Kontorovich, 2013, p. 72)

Tab. 6.6.: Problem-posing situations on the spectrum of unstructured situations.

situations can clearly be separated from unstructured situations by the initial problem that structured situations are based on. Combining the categories free and semi-structured may in future prevent characteristically identical situations from being categorized differently. This different coding of indistinguishable situations may impede the interpretation and comparison of study results, which our proposal aims to prevent.

6.6 Analysis 3: Is it a routine or a non-routine problem?

In Analysis 2, 32 situations were coded as structured. We evaluated the initial problems of these situations with regard to whether they are routine or non-routine problems. Free and semi-structured situations are not considered for this analysis as they do not include initial problems that can be examined. For the situations in Table 6.7 on the facing page, the shopping shelf in situation 12 comes along with three routine problems for which simple addition and subtraction are needed. To solve the initial problem of situation 13, the participants must apply strategies since no known or obvious approach to solve the problem is apparent, thus, it is a non-routine problem for them.

As stated earlier, it depends on the individual whether a problem is a routine or a non-routine, but extreme cases are clearly recognizable. Thus, since it is not possible to decide with certainty whether the initial problems are routine or non-routine without observing the problem-solving process, we refrained from coding them in this regard.

6.6.1 Discussion of Analysis 3: Routine and non-routine initial problems

We argue that the initial problems of a structured situation might influence the resulting problems in the subsequent problem posing with regard to whether mainly routine or non-routine problems are posed. This is relevant because posed problems usually are or should be solved by the participants. Since the process of solving routine problems is different from the process of solving non-routine problems (Rott, 2014), it is advisable to consider which type of problem is to be emphasized by the situation.

	Situation
12	

Structured

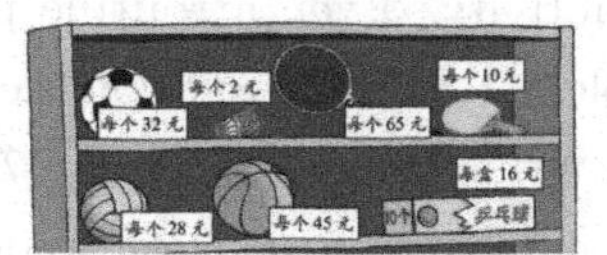

(1) If we want to buy 5 volleyballs, how much do we need to pay?

(2) If we bought three footballs, and paid the cashier 100 dollars, how much can we get for change?

(3) If I want to buy one badminton racket and 10 badminton shuttlecocks, how much do I need to pay?

(4) Please pose two more questions and answer them.

(Jiang & Cai, 2014, p. 396)

13

For the figure [below], one mathematics problem we could ask is: Given that the radius of the smallest circle is one unit, what is the ratio of the area of the largest circle to the area of the smallest circle?

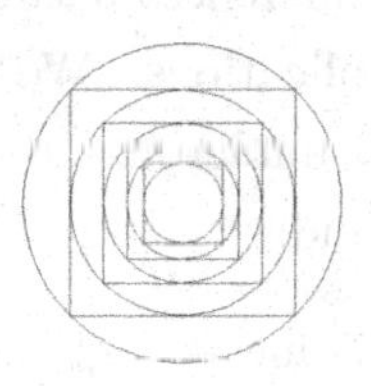

1. Think about how to solve this problem. If you can not [sic] solve this problem, try to come up with a plan or some ideas you have.
2. Pose problems using constraint manipulation or goal manipulation strategy according to the given figure, or the problems you have posed, or any other ideas you have.
3. Could you come up with some ideas or a plan to solve the original problem? If you have solved the initial one, try to solve the one you just posed.

(Xie & Masingila, 2017, p. 116)

Tab. 6.7.: Structured problem-posing situations with a routine (Situation 12) and a non-routine problem (Situation 13) problem as an initial problem.

However, whether the initial problem is routine or non-routine is not sufficient in order to anticipate the resulting problems of a subsequent problem-posing activity. Even a routine problem can lead to posing non-routine problems. Sullivan and Clarke (1991, p. 15), for example, transformed the routine problem *Round 5.77 to the nearest tenth* into a problem which is non-routine for most students: "Describe all numbers that round to 5.8. [...] What is the largest [or smallest] possible number?" The other way around, you can also pose routine problems based on a structured situation with a non-routine initial problem. Therefore, similar to the characterization of problems, it mainly depends on the individual, whether routine problems or non-routine problems are posed.

6.7 Conclusion

There are substantial differences between situations labeled as problem posing in research publications. We strongly believe that a clarified terminology opens up numerous advantages for a research area; as Ruthven (2020, p. 6) said:

> "[I]t seems that an increasingly diverse range of concerns are finding a place under the banner of problem posing so that the original sense of the term – as the substantive formulation of mathematical problems – has been expanded to embrace a spectrum of pedagogical considerations relating to task design, classroom organization, lesson planning and teacher questioning. There is a danger, then, of usage of the term becoming so diffuse as to undermine its analytic power and reduce it to a nebulous slogan."

Therefore, we offer a framework to categorize problem-posing situations. Referring to Ruthven, a well-defined understanding of problem posing serves to obtain the analytical power of the concept of problem posing. However, due to the debatable nature of the topic, our aim is to stimulate reflection and initiate discussion rather than to propose indisputable answers. Table 6.8 summarizes the main points of the three analyses.

Figure 6.3 illustrates the quantitative distribution of the developed framework. It is shown that about one-third of the situations analyzed do not induce a posing activity in the presented understanding. It is also noticeable that more than half of all situations were categorized as unstructured (i.e. free or semi-structured) problem-posing situations. The structured situations, thus, seem to be underrepresented in problem-posing research. This may be critical with regard to the generalizability of study results, since structured problem-posing situations may produce different results. It also shows that only a very small number of situations were available for analysis 3. This identifies an important area of problem-posing research that needs to be expanded. A limitation of the present analyses is that they do not analyse the possibly different activities resulting from the situations. Future studies could deal with this desideratum. In the context of a study, for example, it would be worthwhile examining structured situations with both routine and non-routine problems and analysing the posed problems of the participants with regard to whether they are routine or non-routine problems. Furthermore, similar to problem-solving research, one could investigate whether the process of posing routine problems is different from that of posing non-routine problems.

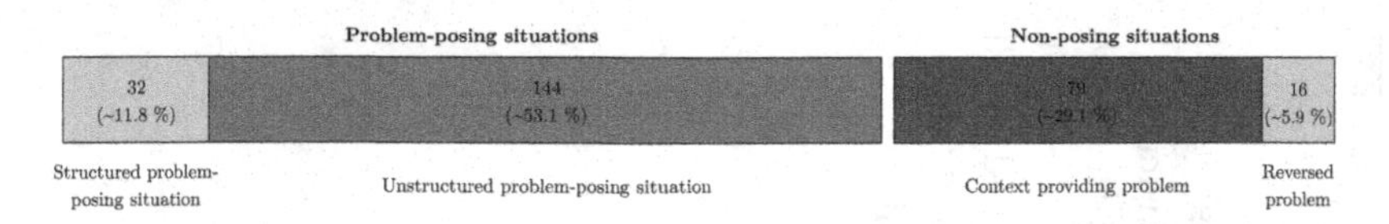

Fig. 6.3.: Quantitative distribution of the situations categorised by the framework.

	Analysis 1	Analysis 2	Analysis 3
Situations Not well-structured problem in the sense that the goal cannot be determined by all given elements and relationships.	**Posing situations** – Situation led to problems that demand a solution – Open for extension – Ill-defined answer	**Structured** Situations that are based on a specific problem (and its solution).	**Routine initial problem** Individual has access to a solution schema for the initial problem.
			Non-routine initial problem Individual has no access to a solution schema for the initial problem.
		Unstructured Spectrum of restrictive-less free situations to semi-structured situations with extensive information that invites exploring structures.	
	Non-posing situations Situations that have been defined as context providing problem and reversed problem.		

Tab. 6.8.: Summary of the three analyses.

The analyses reveal the numerous pitfalls that can occur when selecting problem-posing situations for research and practice. In the following, the implications of the three analyses for research and practice will be presented.

6.7.1 Implications for practice

Analysis 1 separates posing from non-posing situations using openness as a criterion. For practice, we argue that using problem-posing situations with a certain degree of openness conveys students a more authentic picture of mathematics (Silver, 1995). In our opinion, non-posing situations fulfill another, and no less substantial, educational function. The proposal resulting from analysis 2 to consider unstructured situations with varying degrees of restrictions, puts the focus for teachers on how many restrictions they want to provide for their students in problem-posing situations. Analysis 3 is of interest for teachers to consider in advance what type of problems they want their students to pose or to analyze the emerging problems. This does make a difference because problem posing is usually accompanied by problem solving and solving routine problems is different from solving non-routine problems.

6.7.2 Implications for research

Analysis 1 can support researchers in identifying different potential problem-posing situations with regard to their openness. For example, research results from the area of problem posing can be interpreted more specifically: Consider the studies from Cai and Hwang (2002; Situation 1 in Table 6.1) and Arıkan and Ünal (2015; Situation 2 in Table 6.1) in the introduction. The framework introduced here can be used to describe the differences of these studies precisely. Cai and Hwang (2002) use an unstructured problem-posing situation. Arıkan and Ünal (2015) use reversed problems which – based on this framework – are not considered to induce problem-posing activities. Therefore, we argue that it is unproductive to merge these studies when reviewing the current state of research on the relationship between or the influence of problem posing and problem solving because the situations – and thus the activities they provoke – differ considerably in their degree of openness. Also, the criteria for mathematical posing situations provide researchers with a descriptive framework that they can use to classify and justify the selection of situations in their studies.

In Analysis 2, all situations that had not been omitted as non-posing situations in Analysis 1 were further categorized into an adaption of the established categories by Stoyanova and Ellerton (1996). Because we encountered difficulties in distinguishing free and semi-structured problem-posing situations, we proposed that these categories should not be understood as clearly distinguishable, but rather should be placed on a spectrum of unstructured situations that have a different degree of predetermined information. For research, this reduces the chance of an incorrect use of the categories free and semi-structured situations.

In Analysis 3, a closer look is taken at all situations that had been identified as structured in Analysis 2. It is considered whether the initial problem of a structured problem-posing situations is routine or non-routine which may influence the emerging problems of the problem-posing activity. A similar assessment can be made for unstructured situations: Some situations tend to cause routine problems, while others may lead to non-routine problems (Baumanns & Rott, 2019). The analysis of the emerged problems on the spectrum between routine and non-routine problems has been almost completely absent within the reviewed articles. In accordance with our argumentation, Arıkan (2019) stressed the importance of considering the problem type for his studies on problem posing.

These analyses open up a new perspective on established definitions of problem posing and, thus, focus on new aspects that have not yet been considered in problem-posing research. As mentioned in the introduction, the field of problem posing lacks definition and structure. This article's framework is an attempt to close this gap.

7 Developing a framework for characterising problem-posing activities: A review

---

The Version of Record of this manuscript has been published and is available on https://doi.org/10.1080/14794802.2021.1897036. Full reference: Baumanns, L., & Rott, B. (2022). Developing a framework for characterizing problem-posing activities: a review. *Research in Mathematics Education*, *24*(1), 28–50.

L. Baumanns, *Mathematical Problem Posing*, Kölner Beiträge zur Didaktik der Mathematik, https://doi.org/10.1007/978-3-658-39917-7_7

Abstract: *This article aims to develop a framework for the characterisation of problem-posing activities. The framework links three theoretical constructs from research on problem posing, problem solving, and psychology: (1) problem posing as an activity of generating new or reformulating given problems, (2) emerging tasks on the spectrum between routine and non-routine problems, and (3) metacognitive behaviour in problem-posing processes. These dimensions are first conceptualised theoretically. Afterward, the application of these conceptualised dimensions is demonstrated qualitatively using empirical studies on problem posing. Finally, the framework is applied to characterise problem-posing activities within systematically gathered articles from high-ranked journals on mathematics education to identify focal points and under-represented activities in research on problem posing.*

7.1 Introduction

Posing problems has been emphasised as an important mathematical activity by mathematicians (Cantor, 1867; Hadamard, 1945; Halmos, 1980; Lang, 1989) and mathematics educators (Brown & Walter, 2005; Cai & Leikin, 2020; English, 1997; Lee, 2020; Silver, 1994) alike. Since the 1980s, because of its importance, researchers in the field of mathematics education have an increasing interest in investigating the activity of problem posing. As an important companion of problem solving, it can encourage flexible thinking, improve problem-solving skills, and sharpen learners' understanding of mathematical contents (English, 1997). Because of that, problem posing is implemented in the Principles and Standards for School Mathematics (NCTM, 2000). These standards explicitly emphasise problem posing as a genuine mathematical activity (p. 117) which, in combination with problem solving, leads to a more in-depth understanding of the mathematical contents as well as the process of problem solving itself (p. 341).

However, "the field of problem posing is still very diverse and lacks definition and structure" as it "has not been a major focus in mainstream mathematics education research" (Singer et al., 2013, p. 4; see also Cai et al., 2015). In particular, the term problem posing is used to cover numerous activities that differ considerably from each other. This versatility presents difficulties for those who want to better understand problem posing. For this reason, almost 20 years after his seminal contribution to problem posing in 1994, Silver (2013, p. 159) asked: "Is the time ripe for the field to make sharper distinctions among the several manifestations of mathematical problem posing as a phenomenon?" We are convinced that the potential for this discussion exists and want to illustrate the diversity of these manifestations of problem posing referring to activities that may be initiated by the situations in Table 7.1 on the following page. In this article, the term *activity* is intended to refer problem posing in general while *manifestation* refers to a specific characterisation of this activity.

- In situation 1, pre-service primary teachers were asked to generate problems with less restrictions (Tichá & Hošpesová, 2013). The resulting tasks led from routine problems like "Eva has $\frac{1}{2}$ cake, Jirka has $\frac{3}{4}$ cake. How much do they have together?" (p. 138) and more difficult tasks like "$\frac{1}{2}$ kg tomatoes costs [sic] 20 CZK. How much is $\frac{3}{4}$ kg tomatoes?" (p. 138) to non-routine problems like "There are 40 fish in the aquarium. Of it, $\frac{1}{2}$ is red and $\frac{1}{2}$ has a large tail fin. How many specimens of fish are red and how many have a large tail fin? [Are] there any red fish with a large tail fin in the aquarium? If so, how many specimens are there?" (cf. p. 140 f.) However, some of the participants were not aware of what to consider when solving their interesting and substantial problems since no evaluation of their posed problems took place (p. 140).

	Situation
1	Pose three word problems which include fractions $\frac{1}{2}$ and $\frac{3}{4}$. Solve the posed problems. (Tichá & Hošpesová, 2013, p. 137)
2	For the figure [below], one mathematics problem we could ask is: Given that the radius of the smallest circle is one unit, what is the ratio of the area of the largest circle to the area of the smallest circle? 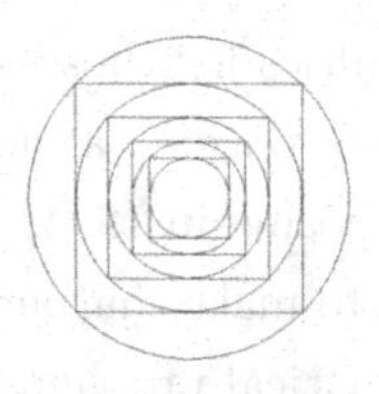 1. Think about how to solve this problem. If you can not [sic] solve this problem, try to come up with a plan or some ideas you have. 2. Pose problems using constraint manipulation or goal manipulation strategy according to the given figure, or the problems you have posed, or any other ideas you have. 3. Could you come up with some ideas or a plan to solve the original problem? If you have solved the initial one, try to solve the one you just posed. (Xie & Masingila, 2017, p. 116)
3	Pose a new problem for a mathematics competition. (cf. Kontorovich & Koichu, 2016)

Tab. 7.1.: Situations that initiate different manifestations of problem posing.

It was only through subsequent joint reflection that they became aware of the complexity of their posed tasks.

- Situation 2 provides an initial non-routine problem, based on which new problems are to be posed by varying its conditions

(Xie & Masingila, 2017). In contrast to situation 1, in which new problems are to be posed by generating tasks without a given initial task, situation 2 focuses on the reformulation of a given task. In this study, problem posing serves as a tool to generate an idea to solve the non-routine initial problem. Thus, the posers are already implicitly asked to evaluate their posed problems in terms of whether they help solve the initial problem.

- Kontorovich and Koichu (2016) (Situation 3) had the chance to observe an expert problem poser preparing tasks for mathematics competitions. From the account of his process, it is apparent that a high amount of evaluative or metacognitive behaviour was involved in his search for non-routine problems. The analysis revealed that new tasks were created through interplay of generation and reformulation – i.e. by retrieving existing knowledge about mathematical content or known tasks that seem suitable for the specific purpose or by modifying conditions of previously generated tasks, respectively.

Each of these descriptions of problem-posing activities refers to the following three dimensions, which will be discussed in this article:

(1) Problem posing as an activity involving the generation of new or reformulation of given problems
(2) Problem posing as an activity involving the posing of routine or non-routing problems
(3) Problem posing as an activity involving regulation of cognition as metacognitive behaviour in problem posing

This article aims to develop a framework that serves to characterise different manifestations of problem posing – as insinuated above – that are characterised through these three dimensions. For this reason, the three dimensions will first be theoretically conceptualised. Afterwards, the application of these conceptualised dimensions is demonstrated qualitatively using empirical studies on problem posing. Finally, the framework is applied quantitively to systematically gathered articles of high-ranked journals on mathematics education to identify focal points and under-represented areas in research on problem-posing activities. This review is driven by the following research questions that structure the analyses:

1. To what extent do the three conceptualised dimensions succeed in characterising different manifestations of problem posing?
2. Which different manifestations of problem posing are represented in high-quality journals on mathematical education?
3. Which focal points and gaps can be identified in research on problem-posing activities?

7.2 What is problem posing?

Before the dimensions of the framework are conceptualised and applied to the reviewed articles, the activity of problem posing is situated within the existing literature.

7.2.1 Definition of problem posing

There are two widespread definitions of problem posing which are used or referred to in a majority of research papers on the topic. The first definition was proposed by Silver (1994, p. 19), who describes problem posing as the generation of new problems and reformulation of given problems. Both activities can occur before, during, or after a problem-solving process. The second definition comes from Stoyanova and Ellerton (1996, p. 518), who refer to problem posing as the "process by which, on the basis of mathematical experience, students construct personal interpretations of concrete situations and formulate them as meaningful mathematical problems." In the following, we adopt the definition of Silver (1994) because the proposed differentiation between generating new and reformulating given problems conceptualises a broad spectrum of problem-posing manifestations which will be considered in the following.

Stoyanova and Ellerton (1996) differentiate problem-posing situations between free, semi-structured, and structured problem-posing situations, depending on their degree of structure. Free situations provoke the activity of posing problems out of a given, naturalistic or constructed situation without any restrictions (see Table 7.1 on page 126, Situation 3). In a semi-structured situation, the problem poser is invited to explore the structure of an open situation by using mathematical knowledge, skills, and concepts of previous mathematical experiences (see Table 7.8, Situation 11). In structured situations, people are asked to pose further problems based on a specific problem, e.g. by varying its conditions (see Table 7.1 on page 126, Situation 2).

7.2.2 Non-posing activities

In empirical studies, there are some situations labelled as *problem posing* (see Table 7.2 on the next page) that we do not consider as problem posing and are excluded in this review (Baumanns & Rott, 2021). We refer to these as *non-posing activities.*

Situation 4 invites providing a context to a given graph. This context can be something like: "The cost of renting a bike is a $ 2 payment plus $ 0.50 for each day you keep it." (Cai et al., 2013, p. 65). We conclude that this task does not necessarily lead to posing a problem that needs to be solved afterward and, therefore, is not considered as problem-posing

	Situation
4	Use the graph below to answer the following questions. 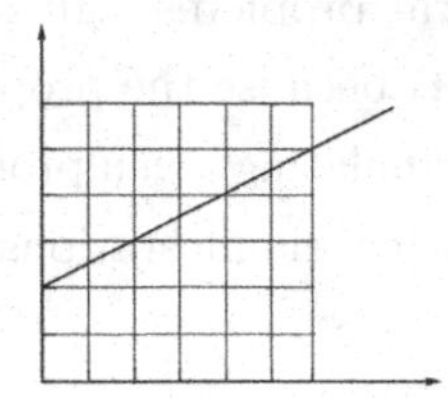 Write an equation that will produce the above graph when x is greater than or equal to zero. Write a real life situation that could be represented by this graph. Be specific. (Cai et al., 2013, p. 63)
5	Write a question to the following story so that the answer to the problem is "75 pounds". "Jason had 150 pounds. His mother gave him some more. After buying a book for 25 pounds he had 200 pounds". (Christou, Mousoulides, Pittalis, Pitta-Pantazi, & Sriraman, 2005, p. 152)

Tab. 7.2.: Situations that induce activities that are not considered as problem posing in this article

activity within this paper. Also, situations in which participants are asked to pose stories, e.g. to open number sentences involving both positive and negative integers like $18+\square=5$ or $-5+\square=21$ (Wessman-Enzinger & Mooney, 2021), have a focus more on providing a context than on posing problems.

In situation 5, students are invited to search for the question to a given context and its answer. The sought-after task, however, is predefined: "How many pounds did Jason get from his mother?" Therefore, after the intended activity, no task can be worked on because its solution is already given. Furthermore, because the situation offers a defined goal, searching for the question that matches the predetermined situation and answer makes working on this situation equivalent to solving a reversed task. These are characterized as mathematical tasks with a defined goal and an undefined question (e.g. Bruder, 2000). Thus, in this article, the activity of reconstructing an implicitly existing question on based on a given solution is not considered as problem-posing activity.

We argue that this demarcation contributes to the further discussion on *problem posing* because they do not apply to the presented understanding of problem posing by Silver (1994). In situation 4, there is no problem that is posed through generation or reformulation. Instead, a context is to be provided. Situation 4 leads to a problem but since the answer to this problem is already been given, the problem is not generated by the means of Silver's definition. Silver (1994, pp. 14–15) stated that the generation of new problems occurs before or after a problem-solving process. Before the question to situation 5 has been posed, there was no problem-solving process and after the question is posed, there follows no problem-solving process since the answer to the problem is already given.

Furthermore, we argue the use of the term problem posing for activities that differ significantly from each other has a negative effect on research (Papadopoulos et al., 2022). The loose use of the term leads to a concealment of the activity and impedes the interpretation of research results. As Ruthven (2020) said:

> "[I]t seems that an increasingly diverse range of concerns are finding a place under the banner of problem posing so that the original sense of the term – as the substantive formulation of mathematical problems – has been expanded to embrace a spectrum of pedagogical considerations relating to task design, classroom organisation, lesson planning and teacher questioning. There is a danger, then, of usage of the term

becoming so diffuse as to undermine its analytic power and reduce it to a nebulous slogan."

Our demarcated understanding of problem posing serves to obtain its analytical power.

7.3 The proposed framework

The proposed framework comprises three dimensions that are visualised as a cube (see Figure 7.1 on the facing page). In the following sections, these three dimensions are conceptualised to develop analytic tools for assessing problem-posing activities.

7.3.1 Dimension 1: Generating and reformulating

As mentioned above, Silver (1994) states that problem posing refers both to (1) the generation of new problems and (2) the reformulation of a given problem.

(1) According to Silver (1994), the generation of new problems can occur *before* or *after* a problem-solving process. Generating before the problem-solving process refers to problem posing based on a given situation in which the goal is the creation of a new problem instead of solving it. Similar to free and semi-structured situations (Stoyanova & Ellerton, 1996), this generation is based on an open situation for which mathematical knowledge, skills, and concepts from previous experiences have to be applied to pose problems.

(2) Reformulating given problems can occur *during* or *after* a problem-solving process. Reformulation mostly refers to the process of varying conditions of a given problem. Schoenfeld (1985b), for

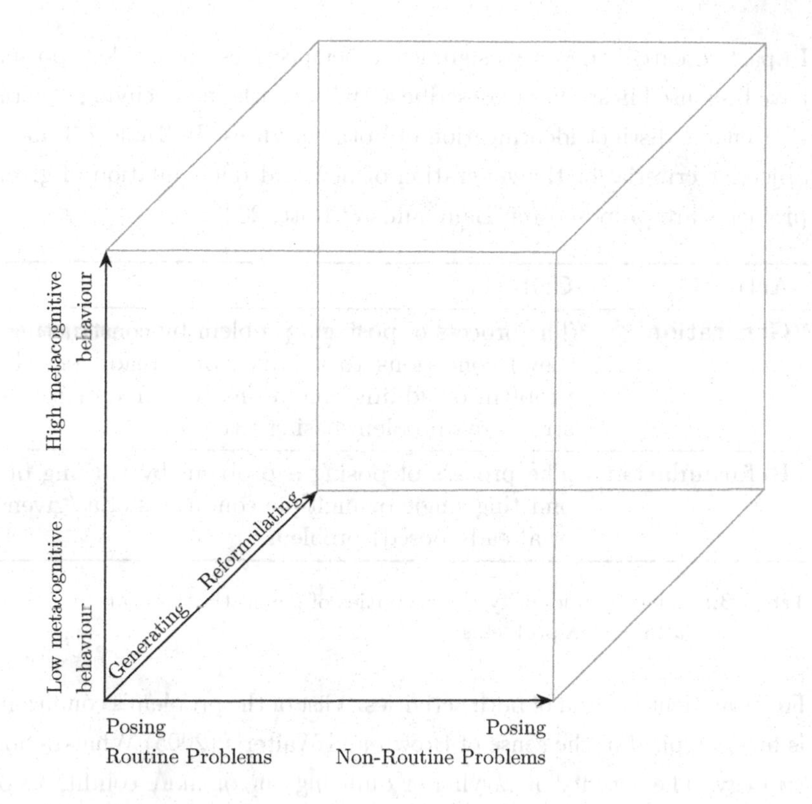

Fig. 7.1.: The proposed framework for assessing manifestations of problem posing

example, suggests posing modified problems by replacing or varying the conditions of a particular problem which is difficult to solve. Posing a problem *after* problem solving refers to Pólya's (1957) phase of "Looking Back" or Brown and Walter's (2005) What-If-Not-strategy in which new and to be solved problems emerge through variations of an initial problem's conditions. In studies on problem posing, this activity is mostly initiated by structured situations (Stoyanova & Ellerton, 1996).

In past research, these two significant perspectives on problem posing have been used in studies to describe activities without specifying criteria that enable distinct identification of both activities. In Table 7.3, more objective criteria for the generation of new and reformulation of given problems are proposed (cf. Baumanns & Rott, 2020).

Activity	Criteria
Generation	The process of posing a problem by constructing (new) conditions to a (given or already posed) problem or adding conditions to a free or semi-structured problem-posing situation.
Reformulation	The process of posing a problem by varying or omitting single or multiple conditions of a (given or already posed) problem.

Tab. 7.3.: Criteria to identify the activities of generating new and reformulating given problems.

For clear identification of both activities, a list of the problem's conditions is first compiled in the sense of Brown and Walter's (2005) What-If-Not-strategy. The activity of varying or omitting one or more conditions of this list is categorised as a *reformulation.* Constructing new conditions to a problem or adding conditions to a free or semi-structured problem-posing situation is categorised as a *generation.* This operationalisation will be applied in the course of the article.

7.3.2 Dimension 2: Routine and non-routine problems

In mathematics education research, the term *problem* has been (and still is being) used in a variety of meanings, leading to some difficulties and misinterpretations of the research literature (Schoenfeld, 1992, p. 337). On the one hand, the term *problem* is used for any kind of mathematical task without differentiation; and on the other hand, the term is used exclusively for non-routine tasks. In this paper, we use the terms *problem* and *task* synonymously. Yet, following Pólya (1966), we specify their meanings with the supplements *routine* and *non-routine.* A problem is a routine problem "if one has ready access to a solution schema" (Schoenfeld, 1985b, p. 74) and a non-routine problem if the individual has no access to a solution schema for a problem. This difference should be exemplified. Integrating a polynomial function of degree 53 can be a tedious computational activity for a mathematician; however, since he or she knows the method of how to integrate a polynomial, this activity is a routine problem for him or her. Otherwise, the problem to determine the number of all squares on a regular chessboard (including squares larger than 1×1) might be a non-routine problem to this mathematician, if the method of how to calculate this number is currently unknown to him/her.

Pólya (1966) emphasises that the ability to distinguish between routine and non-routine problems is important for teachers and researchers. He underlines this with the (certainly not undisputed) statement that "the routine problem has practically no chance, to contribute to the mental development of the student" (p. 126) compared to a non-routine problem. In addition, studies indicated that different beliefs, sub-processes, and heuristics are involved when solving a non-routine problem compared to working on a routine problem (Rott, 2012), which makes this differentiation reasonable for researchers.

In most cases, the decision of whether a given problem is a routine problem or a non-routine problem is evident. Nevertheless, this attribution is specific to the individual; a problem which is a non-routine problem for one person can be a routine problem for another person who knows a solution scheme for this problem Schoenfeld (cf. 1989). Observing a problem-solving process helps to identify with greater certainty whether a particular task is a non-routine problem for a person. Thus, the demarcation between these categories may not be sharp but the extreme cases are clearly recognisable (Pólya, 1966, p. 126). Following this thought, in this article problems are, therefore, identified as representatives on a spectrum between routine and non-routine problems with regard to a specific individual or participants in the respective studies.

7.3.3 Dimension 3: Metacognitive behaviour

According to Flavell (1979), metacognition describes "knowledge and cognition about cognitive phenomena" (p. 906), which roughly means *thinking about thinking*. Based on this understanding and taking a broadly constructivist view of cognition, two facets of metacognition are identified: (1) knowledge about cognition (Cross & Paris, 1988; Kuhn & Dean, 2004; Pintrich, 2002) and (2) regulation of cognition (Schraw & Moshman, 1995; Whitebread et al., 2009).

Knowledge about cognition includes rather declarative knowledge of strategy, task, and person (Pintrich, 2002). Strategic knowledge refers to knowledge about strategies, e.g. when solving problems, and when to apply them. Knowledge of tasks refers to knowing about different degrees of difficulty of tasks and different strategies required to solve them. The person's knowledge includes knowledge about one's own strengths and weaknesses e.g. in problem solving.

Regulation of cognition refers to processes that coordinate cognition. This includes processes of *planning*, *monitoring*, and *evaluation* (Schraw & Moshman, 1995). Planning refers to the identification and selection of appropriate strategies or resources concerning the current endeavour. Monitoring refers to the attention and awareness of the comprehension concerning the current endeavour. Evaluating refers to the assessment of the processes and products of one's learning. As this article investigates activities rather than declarative knowledge, the facet of regulation of cognition will be focused upon.

For research on problem solving, Schoenfeld (1987) already emphasises the activity of *control* or *self-regulation* as the ability to keep track of what someone is doing when solving problems. Building on that, several studies investigate the relationship between metacognitive behaviour and successful problem solving (Lester et al., 1989; Artzt & Armour-Thomas, 1992; Desoete et al., 2001; Kim et al., 2013).

Theoretical considerations on problem posing implicitly contain some aspects of metacognition and metacognitive regulation in particular (Pelczer & Gamboa, 2009; Kontorovich et al., 2012), yet metacognition is rarely explicitly addressed. For example, Voica et al. (2020) found metacognitive behaviour in their study with students as they were able to analyse and reflect on their own posed problems and thinking processes.

Table 7.4 on the following page provides a not-exhaustive list of characteristics that are intended to identify the metacognitive behaviour of regulation of cognition, i.e. planning, monitoring, and evaluating within problem posing. These characteristics are taken from Schraw and Moshman's (1995) descriptions as well as the process-related problem-posing research by Pelczer and Gamboa (2009), and Baumanns and Rott (2020) since the cognitive processes described therein involve metacognitive behaviour while posing problems. In this article, these aspects of metacognitive behaviour will be applied to studies on problem posing.

7.4 Procedure of the literature review

Adopting the guidelines for a systematic literature review by Moher et al. (2009), we systematically gathered articles from all eleven English-language A*-, A-, and B-ranked journals in mathematics education as classified by Törner and Arzarello (2012). These journals are identical to the eleven top-ranked journals in mathematics education research from

Regulation of cognition	Description of behaviour
Planning	• Focus on a starting point of the problem-posing situation to generate new problems • Capturing the conditions and identifying the restrictions of the given problem-posing situation • Reflect necessary knowledge • Express general procedure for problem posing
Monitoring	• Controlling the general procedure for problem posing • Controlling the notation or representation of the posed problems • Assessing consequences on the problem's structure through the modified or new constructed conditions
Evaluating	• Assessing and reflecting on the characteristics of the posed problems (e.g. if it is a routine or a non-routine problem, is it appropriate for a specific target group, is it solvable, interesting, well-defined, etc.) • Reflect on possible modifications of the posed problems

Tab. 7.4.: Identification criteria for regulation of cognition as metacognitive behaviour in problem posing.

the opinion-based study by Williams and Leatham (2017, p. 390). As Williams and Leatham (2017) found out, excluding self-citations, about 80 % of the citations in journals on mathematics educations account to the A*- and A-ranked journals. Therefore, they seem to be perceived by a broad audience and to have a great impact on mathematics education research.

In each of the journals' databases, we searched in all available years from the journal's initiation to 2019 with the search term problem posing which led to 686 articles. Keywords that refer to similar activities, e.g. *question asking* (Mason, 2000), *problem finding* (Jay & Perkins, 1997), or *problem formulating* (Kilpatrick, 1987), were excluded since this review serves to demonstrate the broad spectrum of manifestations that can be found when searching exclusively for *problem posing* as the probably most prominent keyword in research on this activity. Moreover and to our knowledge of the literature, including further keywords would not change the central statements of the article. At the time the review was conducted, 2020 had not yet been completed. Nevertheless, we wanted to include a central additional source from 2020 in the review: all ten articles from the recently published special issue in *Educational Studies in Mathematics* on affects in mathematical problem posing (Cai & Leikin, 2020). These articles were included, because the studies on affects often contained questions that encouraged participants to reflect on their posed problems and the process in general at a metacognitive level (e.g. Bicer et al., 2020), but also because of the influence of ESM articles in our community.[1]

From all these sources, only those articles were extracted that had the term *problem posing* in either their title, abstract, or keywords which led to 84 articles. From the remaining articles, those that did not report on problem-posing activities from participants within an empirical study

[1] A list of all 47 included articles that were part of the review can be found in appendix A.1 on page 277.

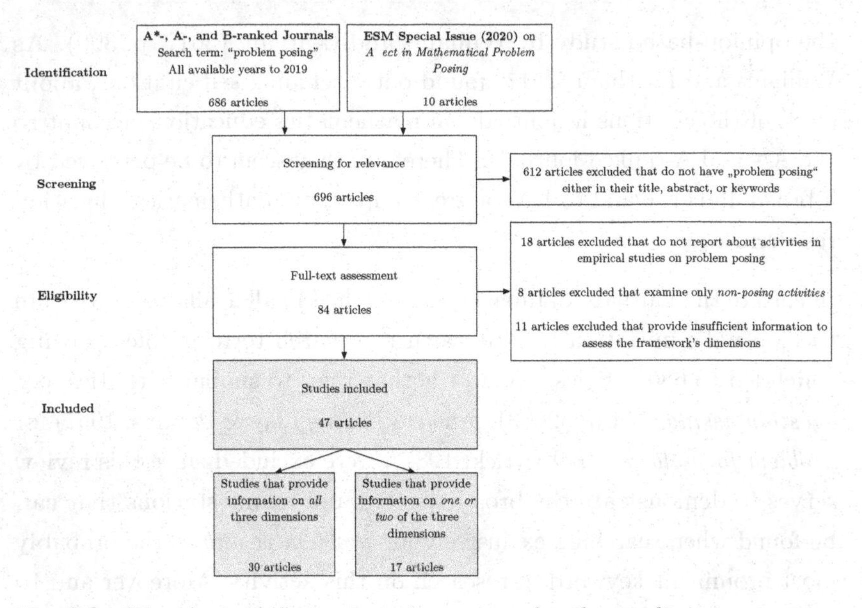

Fig. 7.2.: Flowchart of the systematic process of gathering literature

(e.g. because they discuss problem-posing heuristics from a superordinate subject-matter perspective like Sriraman and Dickman (2017)) were excluded, as the framework was developed for the characterisation of activities. Studies that only examined non-posing activities (as defined and described in section 7.2.2 on page 129) were not included in the review as they induce activities that are characteristically different from posing problems. For the remaining articles, we finally coded those articles that provide sufficient information for an assessment of the framework's three dimensions.

The 47 included articles were coded by the authors. It should be made clear that this coding does not allow us to make unambiguous statements about the manifestations within the reviewed studies as they rarely address the dimensions in detail or provide extensive data material on them. Therefore, our coding is only intended to provide the reader with an overview of studies that contain information on the respective dimensions and their manifestations. Although all three dimensions are to be understood as a spectrum, the articles in the review were coded dichotomously, i.e. generating or reformulating, posing routine or non-routine problems, high or low metacognitive behaviour. This means that we have investigated whether the problem-posing activity of a respective study involves one or both characteristics of each dimension. The dichotomous coding is due to the small number of articles. A more differentiated gradation would, in our opinion, have diluted the results.

For dimension 1, the situations were analysed: unstructured situations are more likely related to generating new problems, structured situations due to the given initial task are more likely related to reformulation. In addition, the processes were considered if, for example, participants first generated a problem and modified it afterwards. In this case, a study was assigned to both characteristics of dimension 1. English (1997), for example, reports that in the problem-posing programme in her study, participants initially discussed important mathematical ideas and "how new problems might be generated from these ideas" (p. 191). Afterwards, they "modified components within given problems to create new problems" (p. 192). So, information on both generating and reformulating is seen within her study and, thus, both characteristics of dimension 1 were coded. This double coding is also possible for the other two dimensions.

For the second dimension, we examined whether the articles provided information about whether posing routine or non-routine problems (for the poser) were the focus of the study. Van Harpen and Presmeg (2013), for example, examined the posed problems with regard to their triviality as to whether they are routine or non-routine problems. Since both task types were common, both characteristics of dimension 2 were coded for this article.

Since nearly none of the reviewed articles explicitly addressed metacognition, we primarily analysed the indications provided by the methodological approach of the studies concerning metacognitive behaviour. Cankoy (2014), for example, encourages his participants in his interlocked problem posing cycle to discuss aspects as sufficiency, quality, or need for modification of the posed problems. Furthermore, we paid attention to general comments on metacognitive behaviour or the absence of it such as the "problems are posed without solving beforehand or deeply understanding the mathematics" or the participants "indicate unawareness of the mathematical potential and scope of [their] problem[s]" (Crespo, 2003, p. 251). Since the article by Crespo, in addition to these statements, also contains information on more reflective behaviour of the participants, for this specific article, high and low metacognitive behaviour was coded for this dimension.

In addition to the authors, a further independent rater coded 9 of the 47 articles ($\sim$ 19.1 %) with regard to all three dimensions. For all three dimensions, agreement was assessed using four codes: (1) information on the first characteristic of the dimension (e.g. generation), (2) information on the second characteristic of the dimension (e.g. reformulation), (3) information on both characteristics of the dimension, or (4) no information on the respective dimension. The inter-rater agreement is substantial (Landis & Koch, 1977) with a Cohen's Kappa of $\kappa = 0.67$. Due to this agreement, the coding of the authors was used for the quantitative evaluation of this review.

7.5 Qualitative application of the framework

In this section, we want to demonstrate the application of the three conceptualised dimensions qualitatively on selected studies that are particularly appropriate to illustrate the spectrum of each dimension.

7.5.1 Dimension 1: Generating and reformulating

Voica and Singer (2013) provide a structured situation (following Stoyanova & Ellerton, 1996) in which fourth to sixth graders with above-average mathematical attainment were asked to modify an initial problem (Table 7.5, Situation 6). The initial problem consists of (at least) seven conditions: (1) square area, which (2) should be fully covered with objects (in particular tiles), (3) there are tiles in two colours (in particular black and white), (4) the corner-tiles should be black, (5) black tiles should be surrounded by white tiles, (6) the number of black tiles should be maximal, and (7) 25 black tiles are given for the coverage.

	Situation
6	A squared kitchen floor is to be covered with black and white tiles. Tiles should be placed on the floor so that in each corner is a black tile, there are only white tiles around each black tile, and the number of black tiles has to be the biggest possible (in the picture there are two examples of such coverage). How many white tiles are needed if 25 black tiles are used? 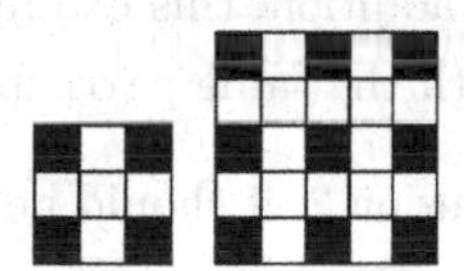 (Voica & Singer, 2013, p. 269)

Tab. 7.5.: Structured problem-posing situation used by Voica and Singer (2013)

Three problems (i–iii) that the participants posed are given in Table 7.6 on page 145. The problems (i) and (ii) may have a different context (e.g. dance show instead of kitchen floor) but the problem's general structure has not changed. Yet, some conditions have been modified or omitted. In problem (i), the 25 coloured tiles of condition (7) was changed to 36 coloured tiles. In problem (ii), the square area of condition

(1) was changed to a rectangular area and condition (7) was, similar to problem (i), changed to 48 coloured tiles. Additionally, condition (6) that the number of coloured tiles has to be maximal, was omitted. Therefore, problems (i) and (ii) have emerged through reformulation. Problem (iii), on the other hand, has emerged through generation because of the construction of new conditions. Although some conditions remained unchanged (e.g. conditions (1) and (2)), new conditions like numbering the pattern of the black fields (consecutive-odd-even-even-odd-consecutive...) were constructed. In summary, structured situations can lead both to processes of generation and reformulation based on the underlying understanding.

Not only structured but also semi-structured or free situations allow both the activity of generating and reformulating. Imagine a poser generates a problem by constructing conditions to a given semi-structured or free situation without a given initial task. Afterward, because the poser may find the posed problem is e.g. not very challenging, interesting, or similar, he/she reformulates it by varying single or multiple conditions of the posed problem. In addition, this example also illustrates that both activities can occur with the same participant in the same situation.

As a transition to dimension 2, it should be mentioned at this point that the posed problems presented above are presumably both routine and non-routine problems. Therefore, regardless of the respective problem-posing activity (i.e. generation or reformulation), both routine and non-routine problems can occur.

7.5.2 Dimension 2: Routine and non-routine problems

The following examples illustrate that the emerging tasks of a problem-posing situation can result on a spectrum between routine and non-routine problems (cf. Baumanns & Rott, 2019). In Table 7.7 on page 146,

	Problems	Conditions {modified} or [omitted]
(i)	On the square wall of the kitchen there are green and blue tiles. They are arranged so that each green piece is surrounded by blue pieces, and the number of green tiles is the biggest possible. The green pieces are in the corners. How many tiles are blue, if 36 are green?	{7}
(ii)	A dance show will take place in a rectangular room. The dancers' costumes will be of two colors: purple and pink. Choreography was made so that in every corner of the scene is one dancer dressed in purple and the purple dancers will be surrounded by only dancers dressed in pink. How many dancers are dressed in pink, if those dressed in purple are 48?	{1}, [6], {7}
(iii)	We have the square next to here. We have to cover it with black in this order: A1, A2,..., A10, D1, D3,..., D9, D2, D4,..., D10, C2, C4,..., C10, C1, C3,..., C9, D1, D2, D3,... 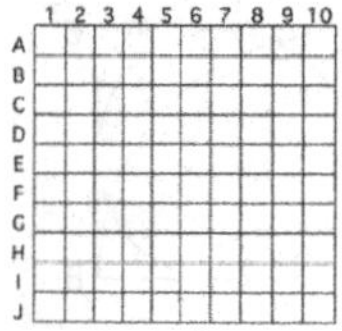 a) What is the 48th covered box? b) When will the G7 box be covered?	

Tab. 7.6.: Problems from participants to the structured situation in Table 7.5 on page 143.

three problem-posing situations are shown to motivate this differentiation. Situations 8 is semi-structured and both situation 9 and 10 are structured situations in the sense of Stoyanova and Ellerton (1996).

	Situation
8	Write a problem based on the following table whose solution would require one addition and one subtraction: **Children** **Bank savings** John 340 Helen 120 Joanne 220 Andrew 110 George 280 (Christou, Mousoulides, Pittalis, Pitta-Pantazi, & Sriraman, 2005, p. 152)
9	The figure contains: the square $ABCD$, the circle inscribed in this square, and the circular arc of centre A and radius AB. Pose as many problems as possible related to this figure, within the time-frame of the next three weeks. Write the posed problems in the order in which they emerged from your mind, if possible, and add a proof, or at least an indication of solving, for each posed problem. Taking into account the above hypotheses and notations, prove that $2 \cdot EC = AC$. 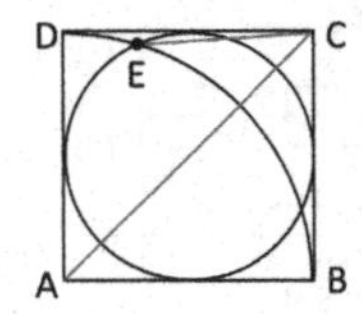 (Singer et al., 2017, p. 39)
10	Round 5.77 to the nearest tenth. (Sullivan & Clarke, 1991, p. 15)

Tab. 7.7.: Problem-posing situations examined in the context of dimension 2

Situation 8 provides a list of bank savings of five people and the restriction to pose a problem whose solution requires one addition and one subtraction. This situation provokes posing tasks like: "Joanne and Andrew combine their savings. How big is the difference between this sum and John's savings?" Tasks like that can be categorised as routine problems for the participants of the study.

Situation 9 already contains an initial problem that is likely to be non-routine for the participants of the study. In the study, it induced the participants to pose tasks like the following ones: Let O be the intersection of the diagonals of the square, F be the second intersection of the full circle and the circle sector ABD, and S be the midpoint of the segment AF. "Prove that $OS = \frac{AO}{2}$. Prove that $\sphericalangle CEO \equiv \sphericalangle CFO$ and that CO is the bisector of angle ECF. Prove that the circle sector OFE is $\frac{\pi R^2 \cdot m(\sphericalangle EOF)}{1440}$" (Singer et al., 2017, p. 45). For most students, these were non-routine problems that the figure provokes to pose.

Situation 10 demonstrates that even initial situations that include routine problems can lead to problems on the whole spectrum between routine and non-routine problems. The following four tasks (cf. Sullivan & Clarke, 1991) can be posed to this situation and serve to demonstrate that spectrum: "Round 5.77 to the next whole number", "Which of the following numbers round to 5.77: 5.764; 5.7649; 5.774; 5.7845", "What is the largest [or smallest] possible number that rounds to 5.8?", and "Describe all numbers that round to 5.8".

If a certain type of task is to be provoked by the problem-posing situation, such a request can be integrated into the situation's text as, for example, Leavy and Hourigan (2020) do: "Choose a class level (first to fourth grade). Make a maths problem that would be a problem for those children." Whether the posed problems actually are routine or non-routine problems the target group the problem is posed for (or the poser him/herself) can only be assumed when there is a subsequent solving process. As already mentioned, it depends on the individual whether a problem is a routine or a non-routine problem. Thus, this assessment can be made with greater certainty, if the situation contains an encouragement to solve the problem. Some studies already use a request to solve the posed task (Xie & Masingila, 2017; Singer et al., 2017).

The posed problems from the situations 8, 9, and 10 show that a similar individual dependency can be drawn for problem posing as for problem solving: Whether a problem-posing situation leads to routine or non-routine problems depends on the respective poser and the purposes the problems are posed for. A teacher who wants their students to train in e.g. the addition of fractions, is likely to produce routine problems for their students. A mathematician who tries to reach a higher level of abstraction concerning a problem by generalising it (cf. Pólya, 1957) is likely to pose non-routine problems for him- or herself.

The analysis of the emerged tasks on the spectrum between routine and non-routine problems has been almost completely absent within the reviewed articles. Only a few studies analyse this explicitly (Crespo, 2003; Van Harpen & Presmeg, 2013; Arıkan, 2019). Such differentiation may be important for educators as well as for researchers in the field of problem posing as the results of such activities differ with regard to their purpose for the students' learning process.

7.5.3 Dimension 3: Metacognitive behaviour

In order to identify the regulative activities of planning, monitoring, and evaluating as metacognitive behaviour, some criteria from Table 7.4 on page 138 are applied to several turns of a transcript excerpt from Kontorovich et al. (2012). This transcript is based on students working on situation 11 from Table 7.8 on page 150.

[1] Shay: I have an idea: if the dimensions of the table were 8×4, after how many hits the ball will enter into pocket B?
[2] Jonny: You mean changing the table's dimensions.
[3] Shay: Yes, and then finding the relationships [between the dimensions of the table and the pocket the ball enters into].
[4] Jonny: So let's check this out.
[5] Sharon: Let's start writing something.
[6] Jonny: We should make a bigger table.
[7] Sharon: I'll do it. What size do you [the group] want?
[8] Jonny: We are given 4×6 and 2×4, so let's make it 8×6.

[9] Sharon: Does everybody agree with 8×6? [Mark and Jonny say "yes"]
[10] Sharon: 8×6 centimetres or squares?
[11] Mark: Squares.
[12] Sharon: Ok [Draws the 8×6 table]. What next?
[13] Mark: Place "A", "B", "C" and "D"
[14] Sharon: Where should I put each of them?
[15] Mark: In the same places as we got.
[16] Jonny [dictates, and Sharon writes]: If the ball made three hits on the sides before entering the pocket in the 6×4 table, and the ball made only one hit on the side in the 2×4 table, how many hits will be in the 8×6 table?
[17] Mark: And which pocket will the ball enter? Harry, what's your opinion: in which pocket it will end up? B?
[18] Jonny: But it's not really a question. We need something more interesting.

In turn 1, Shay chose the starting point of the relation between the dimensions of the table and the number of hits the ball makes until it will enter in pocket B. Jonny makes sure he has understood Shay's statement correctly and the group decides to focus its attention on this context. This activity can be described as planning in terms of cognitive regulation. In turn 8, Jonny refers to the table sizes given by the situation to generate a meaningful modification. This can also be considered a planning activity. The enlargement of the table's dimensions follows from the given dimensions and is, therefore, a controlling procedure of modification. This behaviour is thus interpreted as monitoring. After the problem is posed in turn 16, Jonny evaluates it in turn 18, where he remarks that it is not interesting enough. This can, therefore, be referred to as evaluation.

In summary, the shown example serves as an illustration of the potential to identify metacognitive behaviour when posing problems. High metacognitive behaviour refers to evidence of the criteria listed in Table 7.4 on page 138. Similarly, low metacognitive behaviour refers to the absence of these criteria. The analysis of metacognitive behaviour in problem posing may have a high potential for assessing the quality of the process. An in-depth transcript analysis concerning metacognitive behaviour could gain further insights into this topic.

	Situation
11	Imagine billiard ball Tables like the ones shown below. Suppose a ball is shot at a 45° angle from the lower left corner (A) of the Table. When the ball hits a side of the Table, it bounces off at a 45° angle. In Table 1, the ball travels on a 4×6 Table and ends up in pocket B, after 3 hits on the sides. In Table 2, the ball travels on a 2×4 Table and ends up in pocket B, after 1 hit on the side. In each of the figures shown below, the ball hits the sides several times and then eventually lands in a corner pocket. 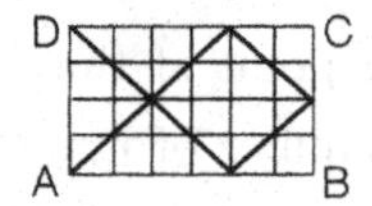 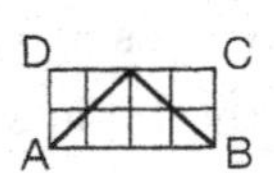 Based on the given situation, pose as many interesting mathematical problems as you can. (Koichu & Kontorovich, 2013, p. 72)

Tab. 7.8.: Semi-structured situation used by Kontorovich et al. (2012)

7.6 Quantitative application of the framework

Following this qualitative insight into the characteristics of the three dimensions, in this section, we present a quantitative evaluation of all articles included in the review. The dichotomous coding of the three dimensions creates eight sub-areas within the framework. These sub-areas are referred to as 3-tuples (Dim. 1, Dim. 2, Dim. 3) whereby Dim. 1 is referred to as *Gen* or *Ref* (*generation* or *reformulation*), Dim. 2. is referred to as *Rou* or *Non-Rou* (*posing routine problems* or *posing non-routine problems*), and Dim. 3 is referred to as *Hi Met* or *Lo Met* (*high metacognitive behaviour* or *low metacognitive behaviour*).

Before conducting the quantitative evaluation, as an example, the article by Martinez-Luaces et al. (2019a) is coded. Concerning dimension 1, they investigate how prospective teachers pose inverse problems of a given non-routine problem through *reformulation.* Although new conditions are occasionally constructed, the basic structure of the initial problem remains unaffected in the reported posed problems. Concerning dimension 2, the problem-posing situation aims at enriching the given non-routine problem and, in fact, the majority of the posed problems were even more sophisticated non-routine problems for participants. Nevertheless, the study also reports on a group that trivialises the initial problem into a routine problem. Therefore, both characteristics of the dimension were coded. Concerning dimension 3, the study indicates that the participants show a high level of aptitude for the difficult task to pose inverse problems. The numerous posed problems reported in the study strongly support the impression that the participants were able to regulate their cognitive process according to the task at hand. In summary, this study was therefore classified in sub-areas (Ref, Rou, Hi Met) and (Ref, Non-Rou, Hi Met) of the framework.

Figure 7.3 on the following page summarises the 30 articles with information on *all* three dimensions in the form of a 3-dimensional heatmap.

Since an article can also have information on both characteristics of a dimension, they can also be in multiple sub-areas. However, there are also 17 articles that do not provide information for *all* dimensions and therefore cannot be placed within the heatmap (see Figure 7.2 on page 140). The assignment of these articles is also given in the Appendix A.2 on page 284. We note that only a small number of articles adequately report on all three dimensions. Accordingly, the developed framework offers a perspective on problem posing that has so far received little attention in studies.

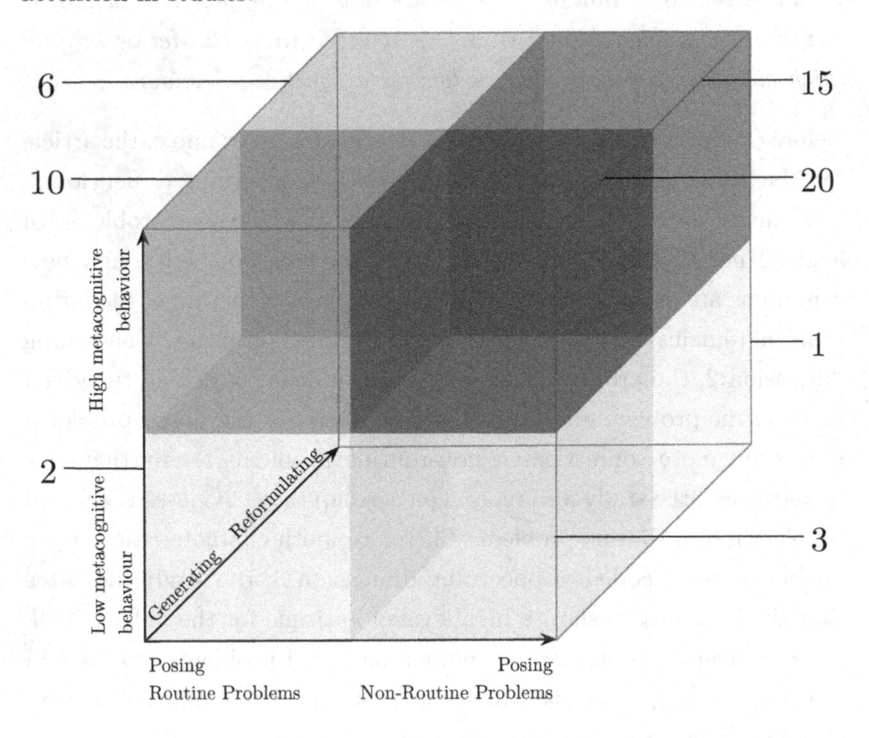

Fig. 7.3.: 3-dimensional heatmap of problem-posing activities reported in studies. The numbers indicate for the number of articles, in which information about the respective part of the cube can be found. No articles were coded as belonging into the only non-visible subarea (Ref, Rou, Lo Met).

The assignment reveals a certain gradient in the number of articles that provide information on different problem-posing manifestations. No article contained information on the manifestation located in sub-area (Ref, Rou, Lo Met). Moving from this corner along the cube's diagonal, the number of articles that provide information on the manifestation located in the corresponding sub-areas increases. As mentioned above, there is little information in the articles about low metacognitive behaviour ((Gen, Rou, Lo Met), (Gen, Non-Rou, Lo Met), & (Ref, Non-Rou, Lo Met)) as studies generally tend not to report too much on less rich problem-posing processes that, thus, show low metacognitive behaviour. Although problem posing is often related to problem solving (as solving non-routine problems), numerous articles provide information on posing routine problems with high metacognitive behaviour ((Gen, Rou, Hi Met) & (Ref, Rou, Hi Met)). A noteworthy majority of articles provides information on posing non-routine problems through generation and reformulation with high metacognitive behaviour ((Gen, Non-Rou, Hi Met) & (Ref, Non-Rou, Hi Met)). This indicates that the most frequent information in the reviewed articles is about the activity that Ruthven (2020, p. 6) describes as the "substantive formulation of mathematical problems".

7.7 Conclusion and discussion

This article highlighted the diversity of problem-posing activities and argued the need for conceptual distinction. As Ruthven (2020, p. 6) points out, there is a "need to develop a precise and powerful notion of problem posing which can command acceptance not only amongst problem posing experts but across the wider mathematics education community." The framework developed in this paper attempts to approach such a notion.

Schoenfeld (2000, p. 646 sqq.) provides eight criteria for evaluating empirical or theoretical work in mathematics education. These criteria are used to discuss the potential and limitation the presented framework offers as a contribution to research and practice on problem posing. However, due to the debatable nature of the topic, we aim to stimulate reflection and initiate discussion rather than to propose indisputable answers. These criteria are (1) descriptive power, (2) explanatory power, (3) scope, (4) predictive power, (5) rigour and specificity, (6) falsifiability, (7) replicability, and (8) multiple sources of evidence.

This framework enables a precise description of the activities induced by the situations in Table 7.1 on page 126 (Criterion 1). In the beginning, these activities were only vaguely outlined. The participants from the study by Tichá and Hošpesová (2013) posed new problems through generation since no initial task was given whose conditions could be modified. The tasks emerged along the entire spectrum between routine and non-routine problems. However, since the participants did not reflect on the tasks that emerged on their own, the respective problem-posing manifestation is located in the sub-areas (Gen, Rou, Lo Met) and (Gen, Non-Rou, Lo Met) (see Figure 7.4 on the facing page).

In the study by Xie and Masingila (2017), problem posing is used as a tool for problem solving. Similarly, Pólya (1957) and Schoenfeld (1985b), among others, also suggest posing a modified problem by replacing or varying the conditions of the initial problem to generate a plan that helps solving it. In this manifestation of problem posing, a better solvable routine or non-routine problem is posed through reformulation on a high level of metacognitive behaviour. Thus, this type of problem posing can be localised by the sub-area (Ref, Rou, Hi Met) and (Ref, Non-Rou, Hi Met) (see Figure 7.4 on the next page).

The expert problem poser in the study by Kontorovich and Koichu (2016) posed new problems both through generation and reformulation.

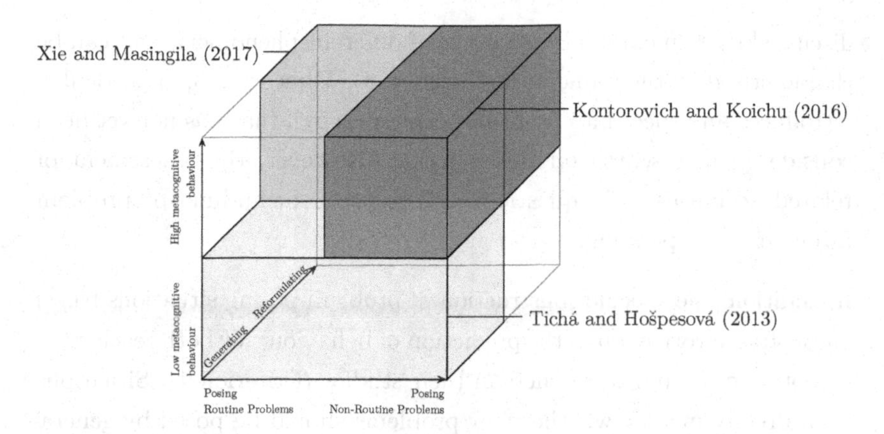

Fig. 7.4.: Localisation of problem-posing manifestations within the three dimensions.

He aimed to pose non-routine problems for a mathematics competition. As the observations of the study show, his problem-posing process is highly reflective. This is why this problem-posing manifestation can be localised by the sub-areas (Gen, Non-Rou, Hi Met) and (Ref, Non-Rou, Hi Met) (see Figure 7.4).

The analyses have also shown to what extent the framework has explanatory power (Criterion 2). On the one hand, it explains the independence of the process, i.e. the problem-posing activity in general (Dimension 1; generating or reformulating), from the product, i.e. the posed problems in a written form (Dimension 2; routine or non-routine problems), when posing problems. On the other hand, metacognitive behaviour (Dimension 3; high or low metacognitive behaviour) fulfils an intermediary function enabling the posers to assess the posed problems concerning the previous process.

Furthermore, the theory developed herein connects three constructs from problem posing (Dimension 1), problem solving (Dimension 2) and psychology (Dimension 3) into one framework. The dimensions discussed therein cover a broad range of different phenomena that can be characterised by the framework (Criterion 3). Dimension 3, in particular, introduces an aspect into problem-posing research that has not yet been considered in research on this activity. However, the assessment of related activities, e.g. question asking, problem finding, or problem formulating, is pending.

In addition, the specific construction of problem-posing situations based on the framework enables the prediction of behaviour for both teachers in school practice and researchers in their studies (Criterion 4). Situations that already include whether new problems should be posed by generation or reformulation, or whether routine or non-routine problems should be posed, allow a more specific prediction of problem-posing activities of students in the classroom or participants in research studies.

For each dimension, the aim was to develop a selective operationalisation (Criterion 5). The reduction of the dimensions to a dichotomy may be abandoned in future considerations as the interpretation of the dimensions as spectra seems to be reasonable. In particular, the framework provides more objective criteria for Silver's (1994) seminal differentiation between the generation of new and reformulation of given problems (Dimension 1). Suggestions were made for the assessment of posed problems through the instruction to solve the problems posed (Dimension 2). Future studies on problem posing may inductively enrich the criteria of metacognitive behaviour within problem posing (Dimension 3). Criteria 4 and 5 also enable the falsifiability of the proposed framework (Criterion 6).

The satisfactory inter-rater agreement shows that different persons, when appropriately trained, identify similar problem-posing manifestation when reading articles (Criterion 7). Since few articles provide deep insight into their data, a more comprehensive application of the framework might better demonstrate its analytical power.

The framework has been successfully applied to all problem-posing studies within the dataset of this review and their very different methodological approaches. It seems worthwhile to examine the framework from different sources. A subsequent stimulated recall interview with the posers could provide further clarity about the framework's descriptions, explanations, and predictions (Criterion 8).

In practice, the developed problem-posing framework enables teachers to trigger specific manifestations of the activity in class. This includes, for example, the conscious decision of whether students should rather pose routine problems for exercise or whether they should understand a non-routine problem in greater depth through variation. For research, the framework identifies further research questions: Do participants pose more routine or non-routine problems to a given situation? Do they get to them by reformulation or generation? To what extent is there a connection between the process and the product? To what extent does the analysis of metacognitive behaviour help to establish this relationship? We believe that in particular, the consideration of metacognitive behaviour in problem posing may be a central enrichment to the field to assess the quality of problem posing and to better understand problem posing in general.

8

The process of problem posing: Development of a descriptive process model of problem posing

The Version of Record of this manuscript has been published and is available on https://doi.org/10.1007/s10649-021-10136-y. Full reference: Baumanns, L., & Rott, B. (2022). The process of problem posing: development of a descriptive phase model of problem posing. *Educational Studies in Mathematics*, *110*, 251–269.

L. Baumanns, *Mathematical Problem Posing*, Kölner Beiträge zur
Didaktik der Mathematik, https://doi.org/10.1007/978-3-658-39917-7_8

Abstract: *The aim of this study is to develop a descriptive phase model for problem-posing activities based on structured situations. For this purpose, 32 task-based interviews with pre-service primary and secondary mathematics teachers working in pairs were conducted who were given two structured problem-posing situations. Through an inductive-deductive category development, five types of activities (situation analysis, variation, generation, problem solving, evaluation) were identified. These activities were coded in so-called episodes, allowing time-covering analyses of the observed processes. Recurring transitions between these episodes were observed by which a descriptive phase model was derived. In addition, coding of the developed episode types was validated for its interrater agreement.*

8.1 Introduction

> *"In re mathematica ars proponendi quaestionem pluris facienda est quam solvendi."* (Cantor, 1867, p. 26)
> Transl.: In mathematics, the art of posing a question is of greater value than solving it.

In his statement, Cantor emphasizes the importance of the ability to pose substantial questions within mathematics. In fact, problem posing is considered a central activity of mathematics (Hadamard, 1945; Halmos, 1980) and at the latest since the 1980s (Brown & Walter, 1983; Butts, 1980; Kilpatrick, 1987), it is being investigated with growing interest by mathematics education researchers. Since the 1990s, it has been widely used to identify or assess mathematical creativity and abilities (Silver, 1994; Silver, 1997; Singer & Voica, 2015; Van Harpen & Sriraman, 2013; Yuan & Sriraman, 2011). Silver (1997, p. 76) emphasizes that to grasp such constructs, both products and processes of problem-posing activities can be considered. However, a strong product orientation within research on problem posing is noticeable Bonotto (2013), Singer et al. (2017), and Van Harpen and Sriraman (2013), that is studies aiming to assess mathematical creativity, for example, often focus on the posed problems rather than the processes that led to them. This is noteworthy since processes are central to educational research. As Freudenthal (1991) states:

> "[T]he use of and the emphasis on *processes* is a *didactical principle.* Indeed, didactics itself is concerned with processes. Most educational research, however, and almost all of it that is based on or related to empirical evidence, focuses on states (or time sequences of states when education is to be viewed as development). States are *products* of previous *processes.* As a matter of fact, *products* of learning are more easily accessible to observation and analysis than are learning *processes* which, on the one hand, explains why researchers prefer to deal with states (or sequences of states), and on the other hand why much of this educational research is didactically pointless." (p. 87, emphases in original)

Although there are studies considering problem-posing processes (Headrick et al., 2020; da Ponte & Henriques, 2013), knowledge about general knowledge about learners' problem-posing processes remains limited (Cai & Leikin, 2020). Only a few studies are dedicated to the development of a phase model for problem posing (Cruz, 2006; Pelczer & Gamboa, 2009). Those models still hold the potential for sufficient generalization and validation. This knowledge could help to develop a more sophisticated process-oriented perspective on problem posing. The few studies that examine the general process of problem posing (Koichu & Kontorovich, 2013; Patáková, 2014; Pelczer & Rodríguez, 2011) may benefit from a validated phase model. Such a model may also be useful for the effective educational use of problem posing in the classroom. This study aims at developing a valid and reliable category system that allows analyzing problem-posing processes. These kinds of conceptual frameworks play a central role in mathematics education research as they enable a better understanding of thinking processes (Lester, 2005; Schoenfeld, 2000).

8.2 Theoretical Background

8.2.1 Problem Posing

There are two widespread definitions of problem posing which are used or referred to in most studies on the topic. As a first definition, Silver (1994, p. 19) describes problem posing as the generation of new problems and reformulation of given problems. Silver continues that both activities can occur before, during, or after a problem-solving process. As a second definition, Stoyanova and Ellerton (1996, p. 218) Stoyanova and Ellerton (1996, p. 218) refer to problem posing as the "process by which, on the basis of mathematical experience, students construct personal interpretations of concrete situations and formulate them as meaningful mathematical problems." In the following, we adopt the definition of Silver (1994) as the differentiation between the activities of generation and reformulation is beneficial for identifying different activities in problem-posing processes. However, both definitions are not disjunctive or contradictory but describe equivalent activities.

In both definitions, the term problem is used for any kind of mathematical task whether it is a routine or a non-routine problem (Pólya, 1966). For the former, "one has ready access to a solution schema" (Schoenfeld, 1985b, p. 74), and for the latter, one has no access to a solution schema. Thus, problem posing can lead to any kind of task on the spectrum between routine and non-routine problems (Baumanns & Rott, 2019; Baumanns & Rott, 2021).

<table>
<tr><th></th><th></th><th>Situation</th></tr>
<tr><td>1</td><td rowspan="2">unstructured</td><td>Pose a problem for a mathematics competition.
(cf. Kontorovich & Koichu, 2016)</td></tr>
<tr><td>2</td><td>Imagine billiard ball tables like the ones shown below. Suppose a ball is shot at a 45° angle from the lower left corner (A) of the Table. When the ball hits a side of the Table, it bounces off at a 45° angle. In each of the examples shown below, the ball hits the sides several times and then eventually lands in a corner pocket. In Example 1, the ball travels on a 6-by-4 table and ends up in pocket D, after 3 hits on the sides. In Example 2, the ball travels on a 4-by-2 table and ends up in pocket B, after 1 hit on the side.
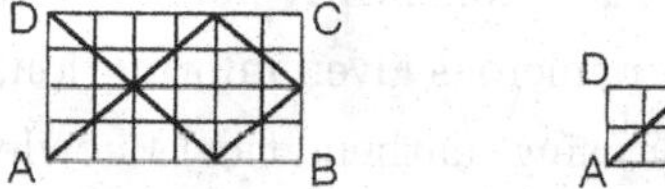

Look at the examples, think about the situation for tables of other sizes, and write down any questions or problems that occur to you.
(Silver, Mamona-Downs, Leung, & Kenney, 1996, p. 297)</td></tr>
<tr><td>3</td><td>structured</td><td>For the figure [below], one mathematics problem we could ask is: Given that the radius of the smallest circle is one unit, what is the ratio of the area of the largest circle to the area of the smallest circle?
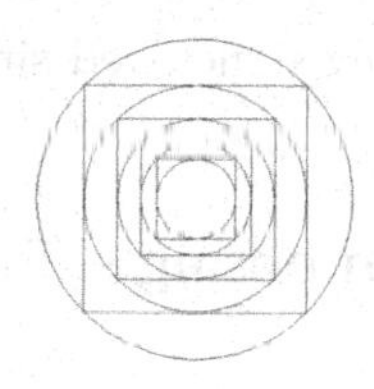
1. Think about how to solve this problem. If you can not [sic] solve this problem, try to come up with a plan or some ideas you have.
2. Pose problems using constraint manipulation or goal manipulation strategy according to the given figure, or the problems you have posed, or any other ideas you have.
3. Could you come up with some ideas or a plan to solve the original problem? If you have solved the initial one, try to solve the one you just posed.
(Xie & Masingila, 2017, p. 116)</td></tr>
</table>

Tab. 8.1.: Unstructured and structured problem-posing situations

Stoyanova and Ellerton (1996) distinguish between free, semi-structured, and structured problem-posing situations depending on the degree of structure. A situation is an ill-structured problem in the sense that its goal cannot be determined by all given elements and relationships (Stoyanova, 1997). Because this study focuses on structured situations and Baumanns and Rott (2021) encountered difficulties in distinguishing free and semi-structured situations, in this article, we distinguish between unstructured and structured situations. Unstructured situations form a spectrum of situations without an initial problem. The given information of these situations reaches from nearly none (see Table 8.1, Situation 1) to open situations with numerous given information, the structure of which must be explored by using mathematical knowledge and mathematical concepts (see Table 8.1, Situation 2). In structured situations, people are asked to pose further problems based on a specific problem, for example by varying its conditions (see Table 8.1, Situation 3). The phase model developed in this article aims at describing problem-posing activities that are induced by situations like those in Table 8.1. In particular, the model is developed using structured situations.

8.2.2 Process of problem posing – State of research

Because products may be more accessible by analysis than processes (Freudenthal, 1991), most problem-posing studies focus on posed problems (Bicer et al., 2020; Van Harpen & Presmeg, 2013; Yuan & Sriraman, 2011). However, consideration of the processes increases in recent studies (Cai & Leikin, 2020; Crespo & Harper, 2020; Headrick et al., 2020; Koichu & Kontorovich, 2013; Patáková, 2014; Pelczer & Rodríguez, 2011). da Ponte and Henriques (2013), for example, examine the problem-posing process in investigation tasks among university students and found that problem posing and problem solving complement each other in generalizing or specifying conjectures to obtain more general knowledge about the mathematics contents. Christou, Mousoulides, Pittalis, Pitta-Pantazi, and Sriraman (2005) describe four thinking processes that occur within problem posing, namely editing, selecting, comprehending/organizing, and translating quantitative information. They found the most able students are characterized through editing and selecting processes. However, compared to the present study, these activities do not tend to describe problem-posing processes by phases. Instead, Christou, Mousoulides, Pittalis, Pitta-Pantazi, and Sriraman (2005) intend to characterize thinking processes in problem posing. Cifarelli and Cai (2005) include problem posing in their model to describe the structure of mathematical exploration in open-ended problem situations. They identify a recursive process in which reflection on problem's solutions serves as the source of new problems.

The studies cited above differ from the present study as follows: They describe and analyze only individual processes, they describe the problem-posing process in terms of thinking processes rather than phases, or they consider problem-posing processes as a sub-phase of a superordinate process. However, there is a lack of studies that attempt to derive a general, descriptive phase model of observed problem-posing processes themselves from numerous processes. For problem-solving research, the analysis of processes through phase models has been established at least since Pólya's (1957) and Schoenfeld's (1985b) seminal works. While their models are *normative*, which means they function as advice on how to solve problems, newer empirical studies on the process of problem solving develop and investigate *descriptive* models, which means they portray how problems are actually solved by participants (Artzt & Armour-Thomas, 1992; Rott et al., 2021; Yimer & Ellerton, 2010). This study also focuses on descriptive models.

Some researchers interpret problem posing as a problem-solving activity (Arıkan & Ünal, 2015; Kontorovich et al., 2012; Silver, 1995) and there are several established models of problem-solving processes (e.g.

Mason et al., 1982; Pólya, 1957). Therefore, it is a reasonable question whether a separate phase model for problem posing is needed. From the observations of problem-solving and problem-posing activities within the present study, we share the argument by Pelczer and Gamboa (2009) that the cognitive processes involved in problem posing are of their own nature and cannot be adequately described by the phase models of problem solving. For problem posing, Cai et al. (2015) state, "there is not yet a general problem-posing analogue to well-established general frameworks for problem solving such as Polya's (1957) four steps" (p. 14).

To find existing research on this topic, we conducted a systematic literature review (Baumanns & Rott, 2021; Baumanns & Rott, 2022a). This review encompassed articles from the high-ranked journals of mathematics education, the Web of Science, PME proceedings, the 2013 and 2020 special issues in *Educational Studies of Mathematics*, the 2020 special issue in *International Journal of Educational Research*, and two edited books on problem posing (Felmer et al., 2016; Singer et al., 2015) From all reviewed articles, three were dedicated to the development of general phases in problem posing similarly to the present study.

Cruz (2006) postulates a phase model based on a training program for teachers (see Figure 8.1 on the next page). For this reason, this phase model is preceded by educative needs and goals. Once a concrete teaching goal has been set (1), the episode type of *problem formulating* begins (2). This episode has a problem as its output which is then *solved* (3). If it cannot be solved, the problem may have to be reformulated (4). A solvable problem is further developed in the episode type *problem improving* (5). The complexity of the problem is adapted to the learning group and compared with the goal (6 and 7). If the comparison shows that the problem is not suitable, either further changes are made to the task (8) or the task is rejected as unsuitable.

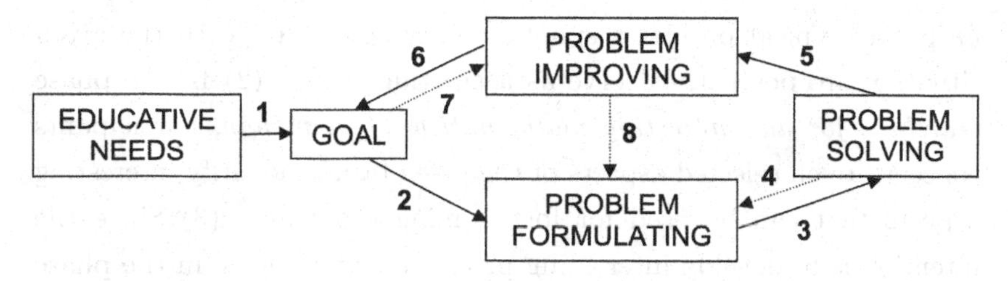

Fig. 8.1.: Phase model of problem posing by Cruz (2006)

Pelczer and Gamboa (2009) distinguish five phases – *setup*, *transformation*, *formulation*, *evaluation*, and *final assessment* – based on the analysis of problem-posing processes in unstructured situations. The *setup* includes the definition of the mathematical context of a situation and the reflection on the knowledge needed to understand the situation. This assessment serves as a starting point for the subsequent process. During the *transformation*, the conditions of a problem are analyzed and possibilities for modification are identified, reflected, and executed. In the *formulation*, all activities related to the formulation of a task are summarized. This includes the consideration of different possible formulations of the problem as well as an evaluation of these formulations. In the *evaluation*, a posed problem is assessed in terms of various aspects, for example, whether it fulfills the initial conditions or further modifications are needed. In the *final assessment*, the process of posing a problem is reflected upon and the problem itself is evaluated, for example in terms of difficulty and interest. In their study, Pelczer and Gamboa (2009) compare experts' and novices' problem-posing processes, identifying different trajectories, that is transitions between the stages. While experts more often go through recursive processes, processes of novices are more linear and often occur without transformation and final assessment.

Koichu and Kontorovich (2013) developed four stages, observed in the context of two successful problem-posing activities: (1) In the *warming-*

up phase, typical problems spontaneously associated with the given situation are posed that serve as a starting point. (2) In the phase *searching for an interesting mathematical phenomenon*, participants concentrate on selected aspects of the given task to identify interesting aspects that can be used for forthcoming problems. (3) Since the intention is to develop interesting problem formulations, in the phase *hiding the problem-posing process in the problem formulation*, the posers try to disguise to the potential solvers in which way the task was created. (4) Finally, in the *reviewing* phase, the posers evaluate the problems based on individual criteria such as the degree of difficulty or appropriateness for a specific target group.

In general, Cruz' (2006) phase model does not allow for sufficient generalization to processes of sample groups that do not pursue school learning goals such as students or mathematicians. The model by Pelczer and Gamboa (2009) has the potential to verify the validity by checking objective coding. The stages by Koichu and Kontorovich (2013) are developed on a small sample of two people and therefore need to be tested for applicability to larger sample groups. All these potentials will be addressed in this article. In addition, although the models presented have certain similarities, they also show numerous characteristic differences. In comparison, phase models for problem solving (Artzt & Armour-Thomas, 1992; Pólya, 1957; Rott et al., 2021; Schoenfeld, 1985b; Yimer & Ellerton, 2010) share a very similar core structure. Thus, there is a conceptual and empirical need for a generally applicable model for problem-posing research.

The need for developing a phase model for problem posing is, furthermore, based on our general observation that the quality of the posed problems did not always match the quality of the observed activity. In our opinion, it is therefore not enough to consider only the products when, for example, problem posing is used to assess mathematical creativity. Furthermore, developing a process-oriented framework serves as research for discussing and analyzing these processes (Fernandez et al., 1994, p. 196).

8.2.3 Research questions

The research goal of this study is to develop a descriptive phase model for the problem-posing activities based on structured situations. The lack of phase models constitutes a desideratum from which the following research questions emerge:

(1) Which recurring and distinguishable activities can be identified when dealing with structured problem-posing situations?

(2) What is the general structure (i.e., sequence of distinguishable activities) of the observed processes from which a descriptive phase model may be derived?

The goal of these research questions is to develop a descriptive phase model that allows analyzing problem-posing processes. To evaluate the quality of this model, we draw on the criteria by Schoenfeld (2000) that can be used for evaluating models in mathematics education. As this type of coding is highly inferential (Rott et al., 2021; Schoenfeld, 1985b), special emphasis is given to interrater agreement.

8.3 The study

8.3.1 Data collection

The present study is a generative study that aims to "generate new observation categories and new elements of a theoretical model in the form of descriptions of mental structures or processes that explain the data" (Clement, 2000, p. 557). For such studies, a less structured, qualitative approach is appropriate that is open to unexpected findings (Döring & Bortz, 2016, p. 192), such as task-based interviews. Task-based interviews have particularly been used in problem-solving research to gain insights into the cognitive processes of participants (Konrad, 2010, p. 482). The interviews were conducted in pairs to create a more natural communication situation and eliminate the constructed pressure to produce something mathematical for the researcher (Schoenfeld, 1985a, p. 178). Johnson and Johnson (1999) also underline that cooperative learning groups such as pairs are "windows into students' minds" (p. 213). For this reason, the interviewer avoided intervening in the interaction process.

The interviews were conducted with 64 pre-service primary and secondary mathematics teachers (PST). The PSTs worked in pairs on one of two structured problem-posing situations, either (A) Nim game or (B) Number pyramid, which are presented in Table 2. The participants were informed that both problem solving and problem posing were central. After the initial problem solving, both situations stated: "Based on this task, pose as many mathematical tasks as possible." This open and restriction-free question should stimulate a creative process. A common question of understanding from participants was, using the example of situation (A), whether they should now pose further Nim games or were also allowed to depart from them. This decision was left to the PSTs' creativity.

In total, 15 processes of situation (A) and 17 processes of situation (B), ranging from 9 to 25 minutes, have been recorded and analyzed. The processes ended when no ideas for further problems emerged from the participants. In total, 7h 46min of video material was recorded and analyzed. Thus, the processes had an average length of 14.5min. Four pairs of PSTs each were in the same room under authentic university seminar conditions. A camera was positioned opposite the pairs capturing all the participants' actions. To accustom them to natural communication in front of the camera, short puzzles were performed before problem posing.

	Situations
(1)	**Nim game** There are 20 stones on the table. Two players A and B may alternately remove one or two stones from the table. Whoever makes the last move wins. Can player A, who starts, win safely? Based on this task, pose as many mathematical tasks as possible. (cf. Schupp, 2002, p. 92) 
(2)	**Number pyramid** In the following number pyramid, which number is in 8th place from the right in the 67th line? Based on this task, pose as many mathematical tasks as possible. (cf. Stoyanova, 1997, p. 70) 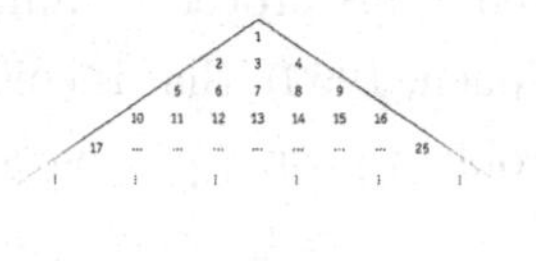

Tab. 8.2.: Structured problem-posing situations used in the study

8.3.2 Data evaluation

For data analysis, we adapted Schoenfeld's (1985b) *Verbal Protocol Analysis*, originally used to analyze problem-solving processes. This method is an event-based sampling. Compared to time-based sampling, the processes are not divided into fixed time segments (e.g., 30 seconds), which are then coded. Instead, new codes are set when the participants' behavior changes. This method has two steps: At first, the recorded interviews are segmented into "macroscopic chunks of consistent behavior" (Schoenfeld, 1985b, p. 292) that are called *episodes* in which "an individual or a problem-solving group is engaged in one large task [.

. . .] or closely related body of tasks in the service of the same goal" (Schoenfeld, 1985b, p. 292). In a second step, the episodes are then characterized in terms of content.

To answer the first research question, verbal protocol analyses were employed in terms of inductive category development (Mayring, 2014, pp. 79–87), meaning that the episode types were developed data-derived. The descriptions of the episode types were additionally concretized in a theory-based manner. For that, the above-mentioned conceptual and empirical findings of problem-posing research (Cruz, 2006; Pelczer & Gamboa, 2009; Silver, 1994), as well as findings of research on phase models in problem solving (Pólya, 1957; Schoenfeld, 1985b), were used. This procedure aims at developing exclusive and exhaustive codes (Cohen, 1960), that is episode types, that can be assigned to the observed problem-posing processes.

To answer the second research question, recurring sequences of the episode types were identified to develop a general phase model. Both general sequences in the observed processes, as well as conceptual insights about problem-posing activities in general, were considered. To analyze the interrater agreement, an independent second coder was trained. At first, the second coder was given the coding manual and a process to code without further comment. For this first coding, cases of doubt were discussed within two hours of training. After this training, the second coder analyzed about 2h 23min of the total video material of 7h 46min which means 10 randomly chosen processes out of 32. Thus, the second coder analyzed about 30.7% of the total video material. Finally, cases of doubt of coding were discussed via consensual validation. These codings were used to calculate the interrater agreement to the author's coding.

The interrater agreement was calculated with the EasyDIAg-algorithm by Holle and Rein (2015). EasyDIAg provides an algorithm that converts two codes of an event-based sampling data set into an agreement table from which Cohen's Kappa (Cohen, 1960) is calculated through an iterative proportional fitting algorithm. Furthermore, in contrast to the classical Cohen's kappa, EasyDIAg provides an interrater agreement score for each value of a category. EasyDIAg considers raters' agreement on segmentation, categorization as well the temporal overlap of the raters' annotations. This makes this algorithm particularly suitable for assessing the interrater agreement of the event-based sampling data set at hand. For the agreement, we used an overlap criterion of 60% as suggested by Holle and Rein (2015). In the online supplement, we provide an example analysis of a process that was coded by the authors and the second rater followed by the calculation of the interrater agreement in this manner.

8.4 Results

First, to retrace the inductive-deductive category development, the problem-posing process of the Nim game by Theresa and Ugur will be described in order to refer back to it when describing the developed episode types. The individual episodes are described without labelling them. The given periods indicate the minutes and seconds (*mm:ss*) of the respective episodes. The recorded time starts with the first attempt at posing problems after the initial problem has been solved. Compared to other participants Theresa and Ugur get the solution of the Nim game quickly and without assistance.

Episode 1 (00:00–00:49): Theresa and Ugur first read the task that should initiate the problem posing. Ugur considers whether new tasks should now be posed in relation to the solution strategy of working backwards. Theresa considers whether the stones should be the focus of new tasks. Afterward, both reflect again on their solution strategy and consider to what extent they can use it for new tasks.

Episode 2 (00:49–02:14): Then other games like *Connect Four* or *Tic-tac-toe*, which may have a winning strategy similar to the Nim game, are collected.

Episode 3 (02:14–05:50): Both participants want to figure out whether there is a winning strategy for Tic-tac-toe. After about three minutes, they assume that an optimal game always results in a draw. They return to the Nim game and ponder whether Player B also has a chance to win safely. They conclude that Player B can only win if Player A does not make the first move according to the winning strategy.

Episode 4 (05:50–07:43): They pose the task of how many stones are necessary for Player B to win safely. Afterward, the text of the task is formulated. They also ask how many moves Player A needs in order to win.

Episo*de 5 (07:43–09:03)*: The last-mentioned question of episode 4 is solved and also generalized. Ugur says, you find the number of moves of Player A by going from the number of stones to the next higher number divisible by three, and then dividing this number by three.

Episode 6 (09:03–09:44): Ugur suggests increasing the number of stones that can be removed from the table. Specifically, he suggests that one to three stones can be removed. Meanwhile, Theresa writes down these ideas.

Episode 7 (09:44–10:32): Theresa writes down the previously posed problems without working on the content of the formulations.

Episode 8 (10:32–13:48): Both play the variation of the Nim game raised in episode 6. They express that they want to develop a winning strategy for this variation. They quickly realize that Player B can safely win the game since multiples of four are now winning numbers and the 20 stones that are on the table at the beginning are already divisible by four. They validate this strategy afterward. At the last minute, the newly posed variation is also evaluated as exciting.

Episode 9 (13:48–14:13): Ugur wants to generalize the game further and poses the task of how to win when the players can remove one to n stones. Theresa asks Ugur if his goal is a general formula.

Episode 10 (14:13–15:48): This task is then solved by Ugur by transferring the structure of the solution of the initial problem to the generalization. Ugur formulates that if you are allowed to remove one to $n - 1$ stones, the player who has the turn must bring the number of stones to n by his turn to win safely.

Episode 11 (15:48–16:42): Subsequently, both work on a suitable formulation for this generalized task.

Episode 12 (16:42–18:28): Theresa notes that solving the initial problem is challenging and therefore suggests providing help for pupils. Theresa suggests that it might help when the pupils first develop a winning strategy for the simple case that the players can only remove one stone. Ugur suggests further help cards which can be requested by the pupils themselves if they get stuck.

Episode 13 (18:28–19:50): Theresa wants to focus on new tasks again. They move away from the initial problem and use the stones to create an iconic representation of the triangular numbers (1, 3, 6, . . .). They formulate the task to find a general formula to calculate the n-th triangular number.

Episode 14 (19:50–21:33): Theresa puts the stones in rows of three so that the structure that leads to the winning strategy is more visible. She evaluates this presentation by emphasizing the usefulness of this method for extensions of the Nim game with more than 20 stones on the table. The process comes to an end as Theresa and Ugur, when asked by the interviewer, agree not to generate any more ideas.

8.4.1 Category development of episode types in problem posing

Using the described evaluation method, five episode categories were developed which allow the observed processes to be described in a time-covering manner. These episode categories are: *Situation Analysis*, *Variation*, *Generation*, *Problem Solving*, and *Evaluation*. In the following, the developed categories of episode types are described. The episodes of the process by Theresa and Ugur (T&U) described above are assigned to these episode types for a better comprehension of the episode types. In addition, we provide further anchor examples in the online supplement. Subsequently, indications are given for coding the individual categories. Finally, the categories are discussed regarding the state of research.

Situation Analysis

Description. During the *situation analysis*, the posers capture single or multiple conditions of the initial task. They usually recognize which conditions are suitable and to what extent, to create a new task by variation (changing or omitting single or multiple conditions) or generation (constructing single or multiple new conditions). In addition, the subsequent investigation of the initial task's solution is summarized in this episode. This also includes the creation of clues or supporting tasks that lead to the solution of the initial task.

In the process of T&U, episode 1 is coded as *situation analysis* as the participants still reflect on their solution strategy. Also, episode 12 is coded as *situation analysis* because both PSTs try to come up with ideas on how to support students with solving the initial problem. A further example of other participants who capture the conditions of the initial problem can be found in the online supplement.

Coding instructions. It is not always clear when the posers are engaged in reading (see *non-content-related episodes* below) or have already moved on to *situation analysis*. Simultaneous coding is possible here. The creation of supporting tasks, which are supposed to assist in solving the initial problem, is interpreted as an analytical examination of the situation and is therefore coded as *situation analysis*.

Variation

Description. During *variation*, single or multiple conditions of the initial task or a task previously posed in the process are changed or omitted. No additional conditions are constructed. In addition, writing down and formulating the respective task is included under this episode.

In the process of T&U, episodes 4, 6, 9, and 11 are coded as *variation*. In episode 6, for example, Ugur varies one specific rule of the Nim game and states that the players are now allowed to remove one to three stones from the table. In episode 9, this is further generalized by variation.

Coding instructions. For the identification of *variation*, the What-If-Not-strategy by Brown and Walter (2005) should be used. The first step of this strategy is intended to extract the conditions of a problem. The Nim game, for example, has at least the following five conditions: (1) 20 stones, (2) two players, (3) alternating moves, (4) one or two stones are removed, and (5) whoever empties the table wins. This analysis should be done before coding. Omitting or varying these analyzed conditions will be coded as *variation*. Also, omitting or varying conditions of a previously posed problem is coded as *variation*.

Generation

Description. During *generation*, tasks are raised by constructing new conditions to the given initial task or a task previously posed in the process. Due to the possible change in the task structure, posers sometimes explain the new task. In addition, writing down and formulating the respective task is summarized under this episode type. Also, free associations, in which tasks similar to the initial task are reminded, are coded as *generation*.

In the process of T&U, episodes 2 and 13 are coded as *generation*. In episode 13, for example, they move further away from the Nim game and use the stones to ask questions about dot patterns.

Coding instructions. The episode types *variation* and *generation* are not always clearly distinguishable from each other. Although the coding focuses on the activity of the poser and not on the emerged task, it can help to examine the characteristics of a task resulting from *variation* or *generation*. In the case of a varied task, the question or the solution structure often remains unchanged. In the case of a generated task, there is usually a fundamentally different task whose solution often requires different strategies.

Problem Solving

Description. *Problem solving* describes the activity in which the posers solve a task that they have previously posed. If a non-routine problem has been posed, the respondents go through a shortened problem-solving process in which the phases of devising and carrying out the plan (Pólya, 1957) are the main focus. In some cases, the posers omit to carry out the plan if the plan already provides sufficient information on the solvability and complexity of the posed problem. If a routine problem has been posed, the solution is usually not explained, since the method of solution is known. However, longer phases of solving routine tasks are also coded as problem solving.

In the process of T&U, episodes 3, 5, 8, and 10 are coded as problem solving as the participants are engaged in solving their posed problems.

Coding instructions. Although solving a routine problem should be differentiated from solving a non-routine problem, both activities are labelled with the same code. However, the commentary of the coding should specify whether an episode is an activity of solving a routine or a non-routine problem.

Evaluation

Description. In the *evaluation*, the posers assess the posed tasks based on individually defined criteria. In the processes observed, posers asked whether the posed problem is solvable, well-defined, similar to the initial task, appropriate for a specific target group, or interesting for themselves to solve it. On the basis of this evaluation, the posed task is then accepted or rejected.

In the process of T&U, episode 14 is coded as *evaluation* and in episode 8, there is a simultaneous coding of *problem solving* and *evaluation*. In episode 8, for example, the participants are initially engaged in problem solving. Towards the end of this episode, they both assess their posed problem based on their interest in solving it.

Coding instructions. Often, *evaluative* statements are made about the course of an episode of *problem solving*, since the criteria for the evaluation of a posed problem (e.g., solvability or interest) are based on sufficient knowledge about the solution of the posed problem. In such cases, the episode types of *problem solving* and *evaluation* cannot be separated empirically, which is why simultaneous coding is permitted. The criterion for this simultaneous coding is that during an episode of *problem solving* an *evaluative* statement must come within a 30-second window for a simultaneous coding to be made. For example, if at least one *evaluative* statement falls during the first 30 seconds of a problem-solving episode, both types of episodes are coded simultaneously. If at least one *evaluative* statement also falls within the following 30 seconds of *problem solving*, both episode types are again encoded simultaneously.

Non-content-related episode types

When participants, for example, ran out of ideas or became distracted during the interview, they engage in the following non-content-related activities. Such activities were also identified in descriptive models of problem solving (Rott et al., 2021). In the process of T&U, episode 7 was coded as *non-content-related episode.*

Reading. The episode of reading consists of reading the situation text as well as a shorter exchange about what has been read to make sure that the text is understood. Since the participants have usually already solved the initial task of the situation, the reading takes place rather in between.

Writing. In the episode of writing, posers write down the text of a problem they have already worked out orally. Also, the posers write down the solution of a previously posed problem. Writing is only coded if no solution or problem formulation is being worked on in terms of content (e.g., specify the problem text).

Organization. Organization includes all activities in which the poser is working on the situation, but where no content-related work is apparent. This includes, for example, the lengthy production of drawings.

Digression. The episode digression is encoded when the posers are not engaged with the situation. This may include informal conversations with the other person about topics that are not related to the task (e.g., weekend activities) or looking out of the window for a long time.

Other. All episodes that cannot be assigned to any other episode type are coded as Other.

Discussion

To provide a theoretical justification of the data-driven episode types of problem posing, we want to connect the five episode types with the presented state of research on problem-posing phase models.

Situation Analysis. In Pelczer and Gamboa's (2009) phase model, we find aspects of *Situation Analysis* in their *transformation* stage. One sub-process of this transformation stage is the analysis of the problem's characteristics. Terminologically, the episode name is based on Schoenfeld's (1985b) analysis, because we observed that, similar to problem solving, posers identify what possibilities for problem posing the given situations provide through their conditions.

Variation. Pelczer and Gamboa (2009) have aspects of *variation* in the stage of *formulation* in which a problem is written down and the formulation is evaluated. *Problem formulating* can also be found in the model by Cruz (2006). The principle of *variation* also plays a central role in problem solving. Schoenfeld (1985b), for example, suggests posing modified problems by replacing or varying the conditions of a particular problem that is difficult to solve.

Generation. Koichu and Kontorovich (2013) consider spontaneously associated problems related to a given problem-posing situation in their model, yet this is only one aspect of the *generation* described above. The distinction between variation and generation is theoretically already conceptualized by Silver (1994). In empirical studies on problem posing, there are so far no objective criteria that enable distinct identification of both activities. The phase model at hand proposes criteria for this distinction.

Problem Solving. Cruz (2006) explicitly mentions problem solving as a stage in his problem-posing phase model. In the model by Pelczer and Gamboa (2009), problem solving is implicit in the evaluation phase, in which the posed problem is assessed and modified. This is presumably done based on the solution of it.

Evaluation. The stage of *evaluation* in the phase model by Pelczer and Gamboa (2009) shares the same name and has similar characteristics. Cruz (2006) implicitly considers *evaluation* when the posers improve the posed problem when they deem it not suitable for a specific learning group. The activity of *evaluation* is closely related to the metacognitive activity of the regulation of cognition (Flavell, 1979; Schraw & Moshman, 1995). In research on problem posing, there are hardly any studies that investigate metacognitive behavior, yet some frameworks implicitly include aspects of it. Kontorovich et al. (2012), for example, consider *aptness* by means of fitness, suitableness, and appropriateness of a posed problem.

8.4.2 Derivation of a descriptive phase model for problem posing

There is no predetermined order of episode types which means there can be transitions from any episode type to any other. However, there is a kind of "natural order" in which episode types appear in most processes and in which transitions *often* occur. This has been indicated by the order in which the episode types were presented in section 8.4.1 on page 176. It was observed that first the conditions of a situation are grasped (*situation analysis*), then new tasks are posed through *variation* or *generation*, these tasks are *solved* in order to *evaluate* them based on the solution. Of course, we did not observe exactly this order in every process, but across the participants and the different problem posing situations, parts of this superordinate pattern were identified. Often the *situation analysis* was observed at the beginning of the process and at the end of a longer phase of *variation*. Also typical were frequent changes between *variation* or *generation* and *problem solving* (sometimes in combination with *evaluation*). Furthermore, problem posing was identified as a cyclical activity. Several participants were observed to revise or to further vary their previously posed problems. Figure 8.2 on the next page shows the T&U's process following Schoenfeld's (1985b) illustrations of problem-solving processes. Several characteristic transitions can be observed in this process. The vertical lines shown in this figure indicate points in time when a new task (either by *variation* or by *generation*) was posed.

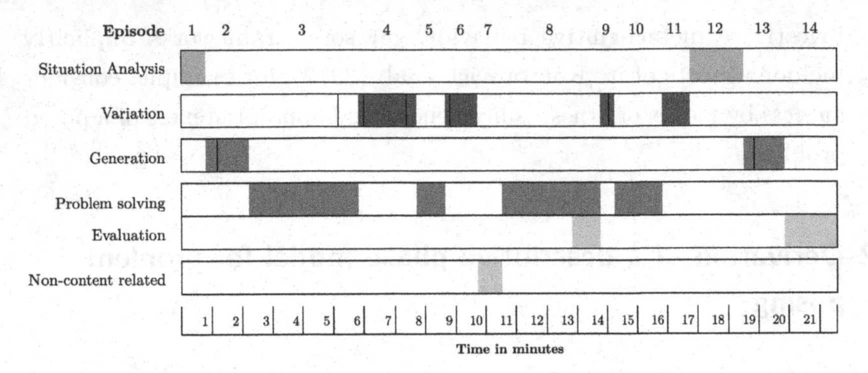

Fig. 8.2.: Example of a time-line chart of the problem-posing process Theresa and Ugur following the illustrations by Schoenfeld (1985b)

From these theoretically justifiable as well as empirically observable patterns in the sequence of episodes, the descriptive phase model shown in Figure 8.3 on the facing page was derived. It contains all five content-related episodes as a complete graph. All transitions indicated by arrows can occur and have been observed empirically in the study. However, not all episode types need to occur in a process. Several participants were observed to revise or to further vary their previously posed problems. In addition, in most cases, not only one but several problems are posed in numerous cycles. The model reflects this observation through its cyclic structure. The model is used to represent all these possible paths within the problem-posing process.

To check the interrater agreement, 30.7% of the total video material of 7h 46min was coded by a second independent rater and combined into an agreement table (see Table 8.3 on page 186) using the EasyDIAg-algorithm (Holle & Rein, 2015). As explained in the results, the episode types of *problem solving* and *evaluation* have empirically often been observed simultaneously, which is why simultaneous coding was allowed. We have, therefore, considered this simultaneous coding as a separate category for the verification of interrater agreement. If the start or end

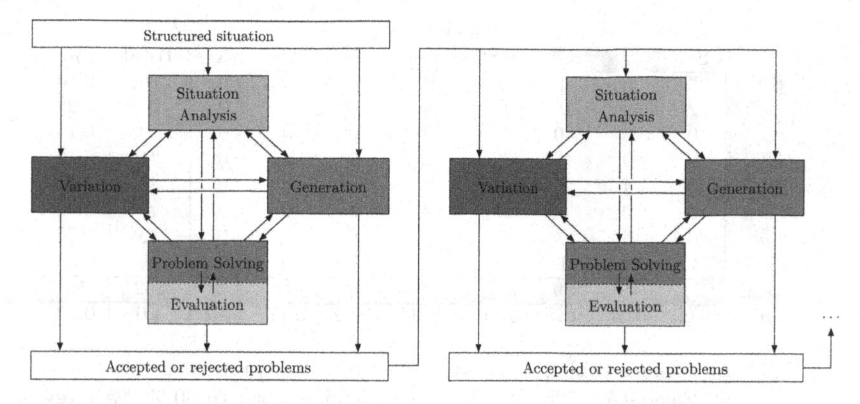

Fig. 8.3.: Descriptive phase model for problem posing based on structured situations

of a process was coded differently in time by the two raters, there are unlinked events in the agreement which are coded as X. The entry X–X in Table 8.3 on the following page can, therefore, not occur empirically.

With a Cohen's Kappa of $\kappa = .81$ the interrater agreement is almost perfect (Landis & Koch, 1977, p. 165). This high level of agreement is particularly gratifying as the evaluation method is a highly subjective and interpretative procedure, yet the developed categories are capable of consistent coding. As anticipated, the biggest coding differences are observed for the categories *variation* and *generation* as well as the distinction between the categories of *problem solving*, *problem solving & evaluation*, and *evaluation*. The Kappa calculated for the separate categories are (with the abbreviations from Table 8.3 as index) $\kappa_{SA} = .87$, $\kappa_V = .83$, $\kappa_G = .72$, $\kappa_{PS} = .87$, $\kappa_{PS/E} = .73$, $\kappa_E = .49$, and $\kappa_O = .97$

8.5 Discussion

This study aimed at developing a valid and reliable model to describe and analyze problem-posing processes. Schoenfeld (2000) provides eight

		Rater 1									
		SA	V	G	PS	PS/E	E	O	X	Total	p
Rater 2	SA	22	0	4	0	0	0	0	0	26	0.10
	V	2	68	0	0	0	0	2	2	74	0.28
	G	0	6	30	4	0	0	0	0	40	0.15
	PS	0	0	0	32	2	0	0	0	34	0.13
	PS/E	0	4	4	2	26	2	0	0	38	0.14
	E	0	0	0	0	2	2	0	0	4	0.02
	O	0	0	0	0	0	0	42	0	42	0.16
	X	0	3	1	0	0	0	0	–	4	0.02
	Total	24	81	39	38	30	4	44	2	262	1.0
	p	0.09	0.31	0.15	0.14	0.11	0.02	0.17	0.01	1.0	**0.81**

Tab. 8.3.: Agreement table for all seven categories of episodes as determined by EasyDIAg. The $\%_{overlap}$ parameter was set to 60 %. (Abbreviations: SA = Situation Analysis, V = Variation, G = Generation, PS = Problem Solving, PS/E = Problem Solving & Evaluation [simultaneous coding], E = Evaluation, O = Others, X = No Match).

criteria for evaluating models in mathematics education: (i) descriptive power, (ii) explanatory power, (iii) scope, (iv) predictive power, (v) rigor and specificity, (vi) falsifiability, (vii) replicability, and (viii) multiple sources of evidence. Criteria (i), (iii), (v), and (vii) will be outlined to discuss the potential and limitations of the presented framework.

Regarding research question (1), five content-related episode types – *situation analysis*, *variation*, *generation*, *problem solving*, and *evaluation* – were identified inductively which enable objective coding through their operationalization. The episode types of the developed phase model enable a specific descriptive perspective on all observed problem-posing processes in the study in a time-covering manner. This description, we argue, provides a better understanding of problem-posing processes in general (i). Furthermore and with regard to research question (2), from the observed processes, a general structure in terms of the sequence of the episodes was identified from which we were able to derive a descriptive process model for problem posing. The high interrater agreement attests to the replicability of the model (vii). The participants of the study were heterogeneous and ranged from PSTs in the first bachelor's semester for primary school to PSTs in the 3rd master's semester for high school. Equally heterogeneous were the processes that could nevertheless be analyzed by the developed model (iii). The detailed descriptions, coding instructions, and theoretical classifications provide specificity to the terms. In the online supplement, anchor examples serve for additional specification (v).

The model developed here provides additional insights compared to existing models (e.g. Cruz, 2006; Pelczer & Gamboa, 2009): It distinguishes the episode types *variation* and *generation* empirically which Silver (1994) already conceptualized theoretically. Additionally, the model encompasses *non-content-related episodes* for the description that have also been identified in descriptive models of problem solving (Rott et al., 2021).

The phase model can now be used to characterize, for example, different degrees of quality of the problem-posing process which is still a recent topic in problem-posing research and for which considering the products *and* processes seems advisable (Kontorovich & Koichu, 2016; Patáková, 2014; Rosli et al., 2013; Singer et al., 2017). Thus, as in problem-solving research (cf. Schoenfeld, 1985b) (cf. Schoenfeld, 1985b), a comparison between experts and novices might be a fruitful approach to identify different types of problem posers. Furthermore and following the process-oriented research on problem solving (Rott et al., 2021), it would be conceivable that the process of posing routine tasks proceeds differently than the process of posing non-routine problems.

Finally, possible limitations to the generalizability of the developed model will be addressed. In general, the model offers one possible perspective on problem-posing processes. Depending on the selected problem-posing situation, sample, or study design, it cannot be ruled out that slightly different or even additional episode types may also occur. We also find other perspectives on problem-posing processes in research (e.g. Headrick et al., 2020). This study considers two specific structured situations with a non-routine initial problem. However, the developed phase model has also been successfully applied to situations with routine initial problems and other mathematical contents within bachelor and master theses. With small changes, the model was also successfully applied to processes based on unstructured situations in several master theses. Moreover, this study has PSTs as a sample. The phase model was successfully applied in bachelor and master theses to other sample groups such as school students and teachers (iii). Therefore, there are strong indications that support the generalizability of the phase model, which could still be clarified in follow-up studies.

9 Identifying metacognitive behavior in problem-posing processes. Development of a framework and a proof of concept

The Version of Record of this manuscript has been published and is available on https://doi.org/10.1007/s10763-022-10297-z. Full reference: Baumanns, L., & Rott, B. (2022). Identifying metacognitive behavior in problem-posing processes. Development of a framework and a proof of concept. *International Journal of Science and Mathematics Education.*

L. Baumanns, *Mathematical Problem Posing*, Kölner Beiträge zur Didaktik der Mathematik, https://doi.org/10.1007/978-3-658-39917-7_9

Abstract: *Insights into the process of mathematical problem posing is a central concern in mathematics education research. However, little is known about regulative or metacognitive behaviors that are essential to understanding this process. In this study, we investigate metacognitive behavior in problem posing. We aim at (1) identifying problem-posing-specific metacognitive behaviors and (2) applying these identified metacognitive behaviors to illustrate differences in problem-posing processes. For these aims, we identified problem-posing-specific metacognitive behaviors of planning, monitoring & control, and evaluating in task-based interviews with primary and secondary pre-service teachers. As a proof of concept, the identified behaviors are applied on two selected transcript fragments to illustrate how a problem-posing-specific framework of metacognitive behavior reveals differences in problem-posing processes.*

9.1 Introduction

In research on problem posing, analyzing the products, that is the posed problems, plays an important role. One reason for this is that the ability to pose problems is often assessed by the posed problems (Van Harpen & Sriraman, 2013; Bonotto, 2013; Singer et al., 2017). However, in our analyses of problem-posing processes, we found that the observable quality of processes can differ, even though the same problems are posed, as will be illustrated with two exemplary processes. One construct that has made this difference within processes tangible for us is *metacognitive behavior*.

At least since Flavell's (1979) seminal work, metacognition has been a central construct of research in psychology and is an established topic of mathematics education research (Schneider & Artelt, 2010). In particular, research on problem solving has benefited from the consideration of metacognitive behavior in the past decades. There are numerous characterizations of metacognitive behavior in problem solving (Garofalo & Lester, 1985; Schoenfeld, 1987; Artzt & Armour-Thomas, 1992; Yimer & Ellerton, 2010) as well as studies on the connection between metacognitive behavior and successful problem solving (Desoete et al., 2001; Özsoy & Ataman, 2009; Rott, 2013; Kuzle, 2013).

Although problem-solving research has benefited greatly from this perspective, it is noteworthy that far less conceptual or empirical research has been conducted in the related field of problem posing. A systematic literature review in high-ranked mathematics education journals has revealed that there are only a few studies that explicitly investigate problem-posing-specific aspects of metacognitive behavior (Baumanns & Rott, 2022a). For example, activities of reflection are observed where problem posers consider solvability or the appropriateness of the posed problem for a specific target group (Pelczer & Gamboa, 2009; Kontorovich et al., 2012). Further, there is a lack of a framework that explicitly addresses the analysis of metacognitive behavior in problem-posing processes. That is in particular striking as researchers often note that the field of problem posing lacks conceptual insights into the activity. Such insights would enable a better analysis and interpretation of the activity itself (Van Harpen & Sriraman, 2013; Ellerton et al., 2015; Ruthven, 2020). Similar to research on problem solving, is can be assumed that the consideration and analysis of problem-posing-specific aspects of metacognitive behavior may be a central enrichment to the field. This can contribute to a better understanding of problem-posing processes.

Based on this research desideratum, this article aims at (1) identifying problem-posing-specific aspects of metacognitive behaviors in problem-posing processes and (2) applying these identified metacognitive behaviors to illustrate differences in problem-posing processes.

9.2 Theoretical Background

9.2.1 Problem Posing

Similar to problem solving, problem posing is considered to be a central activity of mathematicians (Hadamard, 1945; Halmos, 1980; Lang, 1989). Already Pólya (1957) mentioned problem posing as a partial activity in the context of problem solving. Despite this recognition by mathematicians, for a long time, problem posing has received noticeably less attention in mathematics education research than problem solving. At the latest since the 1980s (Butts, 1980; Brown & Walter, 1983; Kilpatrick, 1987), problem posing has been increasingly investigated. In recent years, researchers from mathematics education have shown an increasing interest in investigating and understanding problem posing (Cai et al., 2015; Cai & Hwang, 2020; Cai & Leikin, 2020; Lee, 2020; Silver, 2013). For example, problem posing has been widely used to identify or assess mathematical creativity (Silver, 1997; Yuan & Sriraman, 2011; Van Harpen & Sriraman, 2013; Singer & Voica, 2015; Joklitschke et al., 2019).

There are numerous definitions of problem posing that conceptualize more or less equivalent activities. Silver (1994) defines problem posing as the generation of new problems and reformulation of given problems which can occur before, during, or after a problem-solving process. Stoyanova and Ellerton (1996, p. 218) refer to problem posing as the "process by which, on the basis of mathematical experience, students construct personal interpretations of concrete situations and formulate them as meaningful mathematical problems". Cai and Hwang (2020, p. 2) subsume under problem posing "several related types of activity that entail or support teachers and students formulating (or reformulating) and expressing a problem or task based on a particular context".

Based on the categories by Stoyanova and Ellerton (1996), we distinguish between unstructured and structured problem-posing situations depending on the degree of given information (Baumanns & Rott, 2021). Unstructured situations are characterized by a given naturalistic or constructed situation in which tasks can be posed without or with less restrictions, for example "Consider the following infinite sequence of digits: 123456789101112131415 ... 999100010011002 ... Note that it is made by writing the base ten counting numbers in order. Ask some meaningful questions. Put them in a suitable order" (Stoyanova, 1999, p. 32). To pose meaningful questions, the structure of the situation has to be explored using mathematical knowledge and concepts. In structured situations, people are asked to pose further problems based on a specific problem, for example by varying its conditions. As structured situations are used in this study, examples can be seen in Table 9.2.

9.2.2 Metacognition

Going back all the way to Flavell (1979), who significantly influenced early research on this topic, metacognition is described as "knowledge and cognition about cognitive phenomena", which roughly means *thinking about thinking*. Based on this understanding, two facets of metacognition are distinguished: (1) knowledge about cognition (Cross & Paris, 1988; Pintrich, 2002; Kuhn & Dean, 2004) and (2) regulation of cognition (Schraw & Moshman, 1995; Whitebread et al., 2009).

(1) Knowledge about cognition includes declarative knowledge of strategy, task, and person (Pintrich, 2002). Strategic knowledge refers to knowledge about strategies (e.g. when solving problems) and when to apply them. Knowledge of tasks refers to knowing about different degrees of difficulty of tasks and different strategies required to solve them, for example. Person's knowledge includes knowledge about one's own strengths and weaknesses (e.g. in problem solving).

(2) Regulation of cognition refers to procedural knowledge with regard to processes that coordinate cognition. This facet includes the activities of *planning*, *monitoring*, *control*, and *evaluating* (Pintrich, 2000). Planning refers to setting of a target goal concerning the current endeavor, the activation of prior content knowledge, and activation of metacognitive knowledge, for example knowledge about specific tasks and how to solve them. Monitoring refers to metacognitive awareness and monitoring of cognition, for example when verifying that one has understood the current task. Control refers to the selection and adaption of strategies, for example with the goal to solve a problem. Evaluating refers to activities of reflecting and judging on one's own performance and results in form of an outcome. These activities are not completely distinct from each other and there are different perspectives on their overlap. We follow the perspective of Pintrich (2000), who states that control is mostly conceptualized as dependent or at least highly similar to monitoring (Pintrich, 2000, pp. 459–460). By that, we also follow the approach of Cohors-Fresenborg and Kaune (2007) who summarize monitoring and control under one category (see also Schraw and Moshman, 1995). This is also consistent with past work by Pólya (1957), who indicated roughly these three activities in problem solving even before the construct metacognition was established (Cohors-Fresenborg et al., 2010).

There are attempts to identify the behavior of planning, monitoring & control, and evaluating in learning contexts (Van der Stel et al., 2010; Kaune, 2006). For example, Cohors-Fresenborg and Kaune (2007) provide a category system for classifying teachers' and students' metacognitive activities in class discussions. The main categories of planning, monitoring & control (which they refer to as *monitoring*), and evaluating (which they refer to as *reflecting*) are divided into several sub-categories with different aspects. Table 9.1 on page 196 summarizes the codes for the main categories of planning, monitoring, and evaluating.[1] These codes are used to identify metacognitive behavior by teachers and students through the analysis of verbal protocols of classroom interactions. In addition to metacognitive behavior, Cohors-Fresenborg and Kaune (2007) also consider (negative) discursivity in their coding scheme. Discursivity is understood as a culture in which the teacher as well as the students always refer to each other's expressions, work out differences in approaches and regulate their own understanding. Because it is an additional aspect besides metacognition, and it is specific for classroom interactions which is not analyzed in the study at hand, the aspect of discursivity will not be considered further. In the study at hand, the focus is on the process of problem posing and, therefore, investigates regulation of cognition. In particular, we want to differentiate the activities of *planning*, *monitoring & control*, and *evaluating* for problem posing by adapting the approach by Cohors-Fresenborg and Kaune (2007).

Metacognition is often considered in conjunction with motivation and beliefs. Zimmerman and Moylan (2009) state that proactive self-regulation depends on the presence of motivational beliefs. For example, the metacognitive activity of planning depends highly on aspects like the intrinsic interest into the current endeavor, self-efficacy perceptions, or learning goal orientation. These aspects are deeply interwoven with motivation. It follows that for a holistic view of learners' effort, for example in problem solving, metacognitive processes should be considered in addition to motivational beliefs (Zimmerman & Moylan, 2009).

[1] The current version of this coding manual can be found at https://www.mathematik.uni-osnabrueck.de/fileadmin/didaktik/Projekte_KM/Kategoriensystem_EN.pdf.

Planning		Monitoring		Reflection	
P1	indication of a focus of attention, in particular with regard to tools/methods to be used or (intermediate) results or representations to be achieved	M1	controlling of a subject-specific activity	R1	analysis of structure of a subject-specific expression
		M2	controlling of terminology/vocabulary used for a description/explanation of a concept	R2	reflection on concepts/analogies/metaphors
		M3	controlling of notation/representation	R3	result of reflection expressed by a wilful use of a (subject-specific) representation
		M4	controlling of the validity or adequacy of tools and methods used, in particular with regard to a planned approach or a modelling approach	R4	analysis of the effectiveness and application of subject-specific tools or methods/indication of a tool needed to achieve an intended result
P2	planning metacognitive activities	M5	controlling of (consistency of an) argumentation/statement	R5	analysis of argumentation/reasoning with regard to content-specific or structural aspects
		M6	controlling if the results meet the question	R6	reflection-based assessment or evaluation
		M7	revealing a misconception	R7	analysis of the interplay between representation and conception
		M8	self-monitoring		

Tab. 9.1.: Main categories of *planning*, *monitoring*, and *evaluating* in the category system for classifying teacher and students metacognitive activities in class discussions by Cohors-Fresenborg and Kaune (2007).

9.2.3 Research on Metacognition in Mathematics Education

In mathematics education research, metacognition is of immense importance (for an overview, see Schneider and Artelt, 2010). Assessing metacognitive behavior is often used to investigate the mathematical skills of participants (Mevarech & Fridkin, 2006; Van der Stel et al., 2010). For example, Van der Stel et al. (2010) found out that the quality of metacognitive behavior seems to be a predictor of the mathematical performance in the future by analyzing the regulation of cognition in thinking-aloud protocols of second and third-year students. Most prominently, metacognition is considered in problem-solving research. Considerations on this are mainly based on the ideas of Pólya (1957). Especially Schoenfeld (1985b; 1987; 1992) emphasizes the importance of control or self-regulation in problem solving. He observed that even when students have the content knowledge necessary to solve a problem, a lack of ability to keep track of what they are doing, that is metacognitive behavior, might lead to failure in solving a problem. His theoretical and empirical analyses initiated numerous studies on metacognition in mathematics education – including the present study.

Extensive discussion has been done on characterizing metacognitive behavior in problem solving (Garofalo & Lester, 1985; Schoenfeld, 1985b; Artzt & Armour-Thomas, 1992; Yimer & Ellerton, 2010). For example, Artzt and Armour-Thomas (1992) and Schoenfeld (1985b) both identify the *analysis* of the problem and *planning* the solution as predominantly metacognitive behavior. In the analysis, selecting an appropriate way to reformulate the given problem, for example in order to make it simpler, is referred to as metacognitive behavior. The phase of *planning* is also metacognitive by nature as it involves controlling and self-regulating the solving process (Garofalo & Lester, 1985). Building on that, several studies investigate the relationship between metacognitive activities and successful problem solving and come to the result that successful problem solving is mostly associated with the presence of metacognitive activities (Desoete et al., 2001; Özsoy & Ataman, 2009; Kim et al., 2013; Kuzle, 2013).

9.2.4 Research on Metacognition in Problem Posing

Due to the relevance of considering metacognitive behavior in problem solving, it can be assumed that the analysis of metacognitive behavior could be similarly relevant in problem posing. However, the potential in this area has not been sufficiently exploited to date. As some researchers interpret problem posing as a problem-solving activity (Silver, 1995; Kontorovich et al., 2012) and since there are several established frameworks of metacognitive behavior in problem solving as described above, it is a reasonable question whether there is even a need for a separate framework for metacognitive behavior in problem posing. We follow the perspective of previous studies (Pelczer & Gamboa, 2009; Cruz, 2006; Baumanns & Rott, 2022b) that argue, based on observations of problem-posing processes, that there are characteristic differences between problem-posing processes and problem-solving processes in general. Therefore, also metacognitive processes involved in problem posing differ from those in problem solving. Cognitive and metacognitive processes involved in problem posing seem to be of their own nature for which metacognitive frameworks for problem solving can only serve as a stimulus. From this derives the interest addressed in this study to investigate these problem-posing-specific aspects of metacognitive behavior.

Problem posing includes different cognitive and metacognitive processes (Christou, Mousoulides, Pittalis, Pitta-Pantazi, & Sriraman, 2005; Pelczer & Gamboa, 2009; Koichu & Kontorovich, 2013; Baumanns & Rott, 2022b) which are indicated in the following informally: Problem posers analyze the given situation, examine which mathematical knowledge could be relevant for this, and possibly look for structures in the situation that may lead to an interesting problem. Then, problems are posed and suitable representations of them are sought. The task may then be solved and while solving, the posers may reflect on the difficulty of the task, its appropriateness for an intended target group, or the general interest in the solution. These activities are not limited to cognitive processes, but may also contain metacognitive behavior. Cognitive behavior in problem posing is, for example, applying knowledge of previous mathematical experiences to a given problem-posing situation to pose a problem. Metacognitive behavior would be, for example, to attack a problem in order to assess whether the problem is well-defined or solvable at all.

Theoretical considerations on problem posing implicitly contain some aspects of metacognition and metacognitive regulation in particular (Pelczer & Gamboa, 2009; Kontorovich et al., 2012; Singer & Voica, 2015; Carrillo & Cruz, 2016; Ghasempour et al., 2013), yet metacognition is rarely explicitly addressed as the systematic literature review on problem-posing studies by Baumanns and Rott (2022b) has revealed. The following sections address the few studies that review has identified that implicitly or explicitly address metacognitive behavior.

Voica et al. (2020) mention that they found metacognitive behavior in their study with students as they were able to analyze and reflect on their own posed problems and thinking processes which helped them to become aware of their strenghts. Other studies have also pointed to the lack of metacognitive activities from the participants (Crespo, 2003; Tichá & Hošpesová, 2013). For example, Crespo (2003) writes that four of her thirteen participants "posed [problems] without solving beforehand or deeply understanding the mathematics" and they "indicate unawareness of the mathematical potential and scope of [their] problem[s]" (p. 251). Crespo (2003) identifies unawareness in her participants in the sense of a lack of reflection. This can also be interpreted as a lack of metacognitive behavior. Erkan and Kar (2022) were able to observe metacognitive factors in pre-service mathematics teachers, but these factors varied depending on the problem-posing situation. Their participants considered the strengths and weaknesses of their mathematical knowledge in order to write mathematically valid problems.

Pelczer and Gamboa (2009) as well as Kontorovich et al. (2012) specify thoughts on reflection on the posed problems. In their descriptive process model, Pelczer and Gamboa (2009) state that in the phase of *evaluation*, expert problem posers assess their posed problems in terms of various aspects, for example, whether further modifications are needed. In the phase of *final assessment*, the process of posing a problem is reflected upon and the problem itself is evaluated, for example in terms of difficulty and one's own interest in solving it. Similarly, the framework for handling the complexity of problem-posing processes in small groups by Kontorovich et al. (2012) integrates the facet of *individual considerations of aptness*. Consideration of aptness includes, for example, to what extent the problem poser is satisfied with the quality of the posed problem or whether the posed problem is appropriate for a specific group of solvers.

9.3 Research objectives

As the state of research has shown, metacognition has not been an important factor in problem-posing research. Therefore, the aim of this study is to offer a focused perspective on metacognitive behavior in problem posing. This lack of conceptual and empirical insight constitutes a desideratum which leads us to the following research objective:

(1) Development of a framework for identifying problem-posing-specific aspects of metacognitive behavior (i.e. *planning*, *monitoring &*

control, and *evaluating*) in pre-service teachers' problem-posing processes

As a second research objective, we pursue a proof of concept to identify differences in problem-posing processes based on metacognition. The problem-posing-specific metacognitive behaviors developed in research objective (1) will be applied to selected transcript excerpts of two problem-posing processes. This proof of concept is intended to show to what extent the analysis of metacognitive behaviors in problem-posing processes with the framework developed in (1) makes differences regarding metacognitive behavior in problem-posing processes visible.

(2) Application of the framework developed in (1) as a proof of concept to make differences regarding metacognitive behavior in problem-posing processes visible

9.4 Methods

9.4.1 Research design for data collection

There are several ways to assess metacognitive behavior, for example through interviews, stimulated recall, or eye-movement registration (for an overview, see Veenman et al., 2006). A clear distinction exists between off-line methods which are carried out before or after the current endeavor (often via self-report questionnaires), and on-line methods which are carried out during the current endeavor. As on-line methods seem to be more predictive with regard to the learning performance (Veenman et al., 2006), we chose the approach of video-based content analysis on problem-posing processes of pairs of pre-service teacher students. This method is based on the commonly used assumption that we can make interpretive conclusions about participants' cognition and metacognition from their verbal expressions working in small groups (Artzt & Armour-Thomas, 1997; Goos et al., 2002).

	Situations	
(1)	**Nim game** There are 20 stones on the table. Two players A and B may alternately remove one or two stones from the table. Whoever makes the last move wins. Can player A, who starts, win safely? Based on this task, pose as many mathematical tasks as possible. (cf. Schupp, 2002, p. 92)	
(2)	**Number pyramid** In the following number pyramid, which number is in 8th place from the right in the 67th line? Based on this task, pose as many mathematical tasks as possible. (cf. Stoyanova, 1997, p. 70)	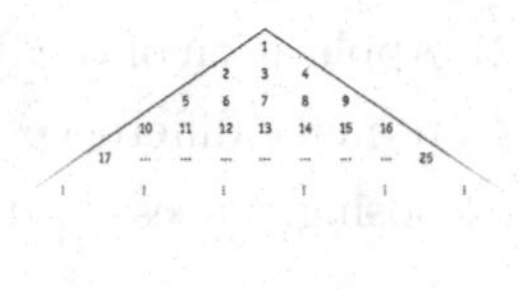

Tab. 9.2.: Structured problem-posing situations used in the study.

The interviews were conducted in pairs in the same room to create a natural communication situation and eliminate the constructed pressure to produce something mathematical for the researcher (Schoenfeld, 1985a, p. 178). Johnson and Johnson (1999) underline that cooperative learning groups such as pairs are "windows into students' minds" (p. 213). For this reason, the interviewer avoided intervening in the interaction process. The participants were asked to speak aloud at any time during interview while posing new problems. The interviews were conducted with 64 pre-service primary and secondary mathematics teachers. 16 students were in the first bachelor semester, 22 in the fifth bachelor semester and 26 in the third master semester. The students worked in pairs on one of two structured problem-posing situations, either (A) Nim game or (B) Number pyramid (see Table 9.2). In total, 15 processes of situation (A) and 17 processes of situation (B) that range from 9 to 25 minutes with an average length of 14.5 min have been recorded. The processes ended when no ideas for further problems emerged from the participants. In total, 7h 46min of video material was recorded. Four pairs of students each were in the same room under authentic university seminar conditions. A camera was positioned opposite the pairs in order to capture all of the participants' actions. In order to accustom them to a natural communication in front of the camera, short puzzles were performed before solving the initial task and the consecutive problem posing.

9.4.2 Data analysis – Assessment of metacognitive behavior

To answer research question (1), we conducted a qualitative content analysis (Mayring, 2000). The main categories of the metacognitive behavior of *planning* (P), *monitoring & control* (MC), and *evaluating* (E) stem from theoretical considerations on regulation of cognition presented above (Schraw & Moshman, 1995; Whitebread et al., 2009), and more specifically from Cohors-Fresenborg's and Kaune's (2007; see Table 9.1) considerations. Although this framework is developed for analyzing classroom interaction, it has been used successfully in paired problem-solving processes (Rott, 2014). However, because problem posing is a mathematical activity of its own kind in our observations (Baumanns & Rott, 2022b), a problem-posing-specific approach was chosen for the present proof of concept. The individual problem-posing-specific characteristics were obtained data-driven through an inductive category development (Mayring, 2000). The category development had four steps: 1. As the unit of analysis, the statements of the participants on the videotaped problem-posing processes were used for the category development. To identify the statements of *planning*, *monitoring & control*, and *evaluating*, the 32 recorded problem-posing processes were analyzed as follows: For the category of planning, we identified participants' statements of setting a target goal for the current problem-posing situation, activating prior mathematical knowledge that helped posing new problems, or activating metacognitive knowledge in form of knowledge on how to pose new problems in general. For the category of monitoring & control, we identified participants' statements of awareness and monitoring of cognition as well as selecting and adapting problem-posing strategies. For the category of evaluating, we identified participants' statements of assessing their problem-posing process or their products, that is their posed problems (e.g., "This is a good problem because it is not too difficult and it is novel compared to the initial problem."). 2. For each identified statement in step 1, a short description was obtained (e.g., *evaluation of the posed problem based on specific characteristics*). Similar descriptions within the main categories *planning*, *monitoring & control*, and *evaluating* were then clustered into subcategories. 3. Afterwards, the developed subcategories were revised by reanalyzing seven problem-posing processes in which particularly many and different metacognitive behaviors were observed and in which participants expressed particularly many verbalizations regarding the interpretation of metacognitive behavior. This reanalysis was used to further refine the descriptions of the categories. 4. After the categories were specified, all the video material was reviewed again in order to draw attention to possible additional categories. No new categories were found in this last step. The quality of the coding was ensured through consensual validation in team discussions which is a common method of ensuring scientific quality in qualitative research (Flick, 2007).

Only for answering research question (2), selected sections of the videotaped problem-posing processes were transcribed. The framework developed in (1) is applied as a proof on concept in the second part of the paper onto the transcripts of these selected sections of two problem-posing processes. This proof of concept is intended to make differences regarding metacognitive behavior in problem-posing processes visible. In the transcripts, the participants' statements are reproduced verbatim. Important actions of the participants are noted in parentheses for understanding the scene. Filler words that do not affect the content have been removed for readability. For the analysis, the transcripts were first read iteratively in order to obtain a rough understanding of the text and to be able to better integrate finer sections of the text into the overall context of the text. The codes developed in research question (1) are then applied to the transcript. The coding of metacognitive behavior of planning, monitoring & control, and evaluation are color-coded in blue, red and yellow in the style of Kaune (2006; Cohors-Fresenborg and Kaune, 2007) to illustrate the distribution of the main categories. We want to emphasize that the analyses of metacognitive behavior were not linked to the correctness of the (mathematical) content. With a wrong argumentation, metacognitive behavior can be just as visible and evaluated as with a correct argumentation.

9.5 Results

9.5.1 Development of a framework for identifying problem-posing-specific aspects of metacognitive behavior

We first want to establish the identified problem-posing-specific aspects of metacognitive behavior of *planning*, *monitoring & control*, and *evaluating*. To do that, the developed types of metacognitive behavior are described and anchor examples of observed processes with regard to the Nim game and Number pyramid (see Table 9.2) are presented.

Planning

In Table 9.3 on the facing page, the categories of regulation of cognition in terms of *planning* are presented. T_1 and T_2 each represent any participants in the study. In Table 9.3, P1 refers to focussing on a starting point of a given situation from which new problem can be posed. This can be, for example, a certain condition, context or solution structure of the given initial problem. Behavior P2 is reminiscent of the well-known "What-if-not"-strategy (Brown & Walter, 2005), in which a similar activity is suggested before the actual problem posing. Behavior (P3) refers to activities in which participants have partly considered what knowledge they or the potential solvers of a posed problem need to have in order to be able to solve it. This behavior could have a greater significance when confronted with unstructured situations in which the necessary knowledge may not be obvious because no concrete initial problem is given. In fact, the phase of *setup* in the mentioned framework by Pelczer and Gamboa (2009) refers the reflection on the knowledge needed to understand the situation. Their framework is based on unstructured situations. Finally, participants named a general procedure for the upcoming problem-posing process, e.g. first vary the initial task in multiple ways, then solving the varied tasks (P4).

Monitoring

In Table 9.4 on page 209, the categories of regulation of cognition in terms of *monitoring* are presented. MC1 follows on from P4 and characterizes that metacognitive behavior in which participants control the problem-posing process. This occurs, for example, by suggesting to solve a posed problem first before focussing on another one. Controlling the notation or representation of the posed problems (MC2) refers to, for example, figures drawn to illustrate a problem or to the formulation of the specific question so that it becomes understandable and precise. We

Code	Description	Anchor example
P1	Focus on a starting point of the problem-posing situation to generate new problems	(The participants chose one variation of the Nim game at the very beginning of their problem-posing process.) T_1 When you think about a new game, I'm thinking about when you're allowed to remove one to three stones. T_2 Yes or, if you, exactly, if you increase the number of stones you are allowed to remove.
P2	Capturing the conditions and identifying the restrictions of the given problem-posing situation	(The participants are at the end of a longer period of varying the Nim game and are now thinking about what other conditions there are that could be varied.) T_1 Are there any other possibilities, I mean variables that can be influenced in this? This would really be the number of players, number of stones, how many do you remove? T_2 Amount of steps. T_1 Exactly.
P3	Reflect necessary knowledge	(The participants posed a variation of the Nim game in which you are allowed to remove one to three stones. Afterwards, they pose another variation in which you can remove one to four stones. Both recognize a pattern and reflect the mathematical content.) T_1 Actually, [...] that would be a cool introductory assignment to me introducing modular arithmetics. T_2 Yeah. T_1 A little bit at least, right? Or at least a more advanced task.
P4	Express general procedure for problem posing	T_1 Yes, now I can try to vary all sorts of things. For example, I could vary the number of stones at the beginning. [...] Change the number of removable stones per turn or something. Change the number of players. I have no idea yet which of these things will be difficult to say simply trivial.

Tab. 9.3.: Planning activities in problem-posing processes

observed that participants made a modification to the initial problem and analyzed the consequences that this modification had on the newly created problem (MC3), for example for the solution structure or its difficulty. The code MC4 was identified when participants analyzed the mathematical structure of the given situation in order to get to a new problem or analyzed the structure of a posed problem in order to be able the characterize it, for example with regard to its solvability or appropriateness.

Evaluating

In Table 9.5 on page 210, the categories of regulation of cognition in terms of *evaluating* are presented. Assessing and reflecting on the characteristics of a posed problem (E1) was seen when participants posed a problem and got an idea about how to solve it. Then, they often evaluated whether their posed problem is interesting, solvable, or appropriate for a specific target group. This behavior was already mentioned in previous studies on the problem-posing process (Pelczer & Gamboa, 2009; Kontorovich et al., 2012; Baumanns & Rott, 2022b). A reflection on modifications of the posed problems (E2) was observed when the posed problem lacks a specific characteristic, for example it is too easy or too difficult, it is not very interesting, or it is too similar to the initial problem.

9.5.2 Proof of concept: The cases of Tino & Ulrich and Valerie & Wenke

In this section, we will analyze two cases – one by Tino & Ulrich and the other by Valerie & Wenke – with regard to the aspects of metacognitive behavior that have been developed in the previous section. The aim of this section to illustrate different degrees of metacognitive behavior

Code	Description	Anchor example
MC1	Controlling the general procedure for problem posing	(The participants are relatively at the beginning of their problem-posing process with regard to the Number pyramid and initially pose several variations. T_1 asks whether the questions of the initial task should be retained. T_2 suggests as a procedure first tasks with the same question are posed and then further changes can be made.) T_1 Are we going to use the same ... question? T_2 Yes, then we can pose another task with a different one afterwards.
MC2	Controlling the notation or representation of the posed problems	T_1 You could also make a pyramid like this and then start with 5 in the next row. That this is not always the row whose square number it is, but that it is done in steps of 5, so that you first have 5 numbers, then 10. T_2 This is not a pyramid, this is something like this (shows the shape of the resulting figure by sharp movements).
MC3	Assessing consequences for the problem's structure through the modified or new constructed conditions	(The participants play the variation of the Nim game in which the players are allowed to remove 1 or 3 stones from the table.) T_1 So, here I always win T_2 Yes, you always win in this situation. T_1 Just in a different system. T_2 Yes, but the other way round, because now the one who does not start always wins.
MC4	Controlling mathematical activities related to a posed problem	T_1 In which row is the second triplet of prime numbers, or so? T_2 Well, there aren't that many prime triplets, are there? T_1 Aren't there two? There are those right at the beginning. 3, 5 and 7 and then there's another one somewhere. Wasn't that 3... No, it wasn't. T_2 19, 21, 23. Oh no, 21 is not a prime number.

Tab. 9.4.: Monitoring & control activities in problem-posing processes

Code	Description	Anchor example
E1	Assessing and reflecting on the characteristics of the posed problems (e.g. if it is appropriate for a specific target group, solvable, interesting, well-defined, etc.)	(The participants posed the variation where the stones are built up into a square pyramid with 4^2 stones at the bottom, then 3^2, then 2^2 and 1^1 at the top. You can remove 1 or 2 pieces per turn but only from one level.) T_1 But now you have to think, ok do you want ... at the end there are 16 stones, do you want to start in this layer or not? Then there are 9 stones above that, ok. Then you have to decide, ok, if there are 9 stones, do you want to start in this layer or not? But I think that's actually cool. Because then you still have to think about the number of stones in each layer. That's actually a cool game, because you have to apply the rules differently again. T_2 Yes. It's really good. T_1 It's good, right?
E2	Reflect on possible modifications of the posed problems	(T_1 raises the modification that player A may remove 1 to 3 and player B 1 to 2 pieces from the table. They then play through this idea and realize that player A can of course use the same winning strategy. Oskar reflects this as player A and proposes a modification on this basis.) T_2 (laughs) It's cheeky to put them down like that (and points to the triple packs that Oskar divided the pieces into) T_1 (laughs) Yes, that's why I mean that. It would be more interesting if I could start and only take two away. Or three or two, for example.

Tab. 9.5.: Evaluating activities in problem posing

in two problem-posing processes that have a similar product. The two selected fragments are identical in numerous features of the outer structure: All four participants are pre-service high-school teachers in their third semester of their master degree. In the analyzed fragments, they work on posing new problems for the Nim game (see Table 9.2). In the shown fragments, both pairs of participants pose the same problem, namely the variation of the Nim game in which the players are allowed to remove 2 or 3 stones from the table. The transcript analysis is conducted to illustrate the different degree of metacognitive behavior between those two processes in which a largely identical product is produced.

Tino & Ulrich – Analysis of the metacognitive behavior

The following excerpt from the process of Tino and Ulrich takes place in the first half of their process. Beforehand, they already posed, solved and analyzed several new variations of the Nim game such as: What if there are 21 stones on the table in the beginning? What if you could remove 1, 2, or 3 stones from the table? What if you also get a winning point if you have removed more stones from the table than your opponent? Then they pose the problem that you are only allowed to remove 2 or 3 stones from the table. The development of this problem is shown in the following transcribed excerpt that takes 3m 12s. For a better visual assessment of the density of metacognitive behavior, in this transcript, the codes for the metacognitive behavior are already set and marked in color.

1	U:	So and now we make a next variation namely you may no longer take 1 to 3, but you may either take 2 stones or 3 stones.	P1
2	T:	What about the variation with number of stones is also a victory factor?	
3	U:	Oh yes, we can do that too...	

4	T:	At least we can notice for a moment, right?	
5	U:	... But I would like to do that later, I would like to save that for a little, so this is definitely also a variation.	P1
...			
10	T:	(referring back to the problem posed in Turn 1) So you can't just remove one tile, right? (writes down) Okay.	MC2
11	U:	Here is a scenario; at 4 nobody wins (5 sec).	
12	T:	When I take 3.	
13	U:	Yes. This is a new game. I think it's great, it's already very good. There are situations where nobody wins. Yeah, it's like at the game	MC3 & E1
14	T:	Tic-tac-toe.	
...			
17	U:	Yes, I can't remember exactly. Already interesting! It is already interesting.	E1
18	T:	Yeah, it's definitely interesting.	E1
19	U:	On 5...	
20	T:	The question is, the question is whether one still admits that one also loses if one only has one stone left, but one can no longer move.	E2
21	U:	Yes, but I would not do that.	E2
...			
24	T:	Because then you practically keep it up, right? The winning strategy.	MC3
25	U:	But I'd say it's a little lame somehow.	E1
26	T:	Yes, of course, but just to think about how I can keep this system up, it would probably be a possibility.	E2
27	U:	Yes, that's true. Then, exactly, then the system would also be upright. But let's move on to the next step. Now, if you confront the other with 5, you win.	E2 & MC1

In turn 1, Ulrich poses a new variation, in which only 2 or 3 stones may be removed from the table, as starting point. This new starting point is derived from a previous task (1 to 3 pieces may be removed). Since Ulrich sets a new focus for the upcoming problem-posing activity, this statement is coded as *planning* (P1). After Tino has thrown in what happened to one of the previous ideas, Ulrich refocuses on the problem he just posed and says that the task Tino mentions can be dealt with later. Therefore, this statement is coded as *planning* (P1).

In turn 10, Tino tries to find a formulation for the problem that was posed in turn 1. He writes down this task as a negation that one may not just remove one stone from the table. His thinking about the formulation of the question represents a control of the notation or representation of the problem and is therefore coded as *monitoring & control* (MC2).

Ulrich says that this change results in a "new game". He probably means a new kind of outcome of the game, where nobody wins. This assessment of the consequences that their variation has for the Nim game was coded as *monitoring & control* (MC3). Ulrich states that he likes the consequences that follow from their variation since they are different from the initial task. Therefore, this is coded as *evaluation* (E1). In turn 18, Tino agrees with Ulrich's positive evaluation of the game.

Tino interjects in turn 20 whether they should modify the new game due to this situation by adding that a player loses even if s/he can no longer remove stones from the table. Ulrich says that he would not make this change. In both statements, the participants consider to modify the posed problem so that the game has a definite winner. Therefore, statements related to that consideration is coded as *evaluation* (E2).

Tino states in turn 24 that this change would restore the original winning strategy of the initial task. By that, he assesses the consequences of his slight modification and compares it to the initial task. Therefore, this statement is coded as *monitoring & control* (MC3).

Ulrich does not seem to like this change, perhaps because it would bring him too close to the initial task. In turn 13, he seemed to like this new element very much. This statement is coded as *evaluation* (E1).

Tino reflects in turn 26 that one could modify the game with his suggestion in order to maintain the original winning strategy of the initial task. This is a reflection on their modification and, thus, is coded as *evaluation* (E2).

Ulrich initially agrees with Tino's previous assessment (E2). Then, he focuses on a solution strategy of the modified game again and thinks about the situation in which five stones lie on the table. With his statement, he is controlling the process which is why this statement is coded as *monitoring & control* (MC1).

30	T:	Right. If I take 3...	
31	U:	Your goal is 5.	
32	T:	... you have 2 to choose from. If I take 2, you have 3 to choose from.	
33	U:	Your goal is 5 and then it is exactly the same.	
34	T:	Yes.	
35	U:	Then you go .. it is a multiple of 5.	
36	T:	That means, what is the change now? Before it was a multiple of 4, where you had lost. (referring back to a previously posed problem where you can remove 1 to 3 stones)	MC3
37	U:	Exactly, because because you always add the largest plus the smallest. You could have maximum 3 and minimum 1 and here the largest is 3 plus the smallest is 2.	MC3
38	T:	2, yes. What if one could remove 3 or 4?	P1

In turn 36, Tino now assesses the changes in the solution strategy of the new game against the background of the previously posed tasks. In the next tudidacrn, Ulrich contrasts the solution strategy of the new game against the background of the solution strategy of previous game and tries to bring both together under one mathematical thought. As both statements are reflections on the winning strategy of the new game and how it is related to the previous game, they are coded as *monitoring & control* (MC3).

Tino focuses on another variation that seems to result from the above considerations. The background could be that Ulrich's consideration in Turn 37 is to be checked on a similar task. In that way, Tino sets a new focus for the process which is why this statement is coded as *planning* (P1).

Overall, it can be seen that Ulrich has a controlling influence on their process (MC1). In both Turn 5 and Turn 27, he determines the direction in which the process should continue. First, by specifying the focus of the next considerations and then by wanting to better understand the problem that they have posed. Tino's behavior is characterized by the fact that he wants to further modify the game that has been posed (E2) so that the inevitable situation that has arisen can be avoided.

Valerie & Wenke – Analysis of the metacognitive behavior

The following excerpt from the process of Valerie and Wenke takes place quite early in the process. Beforehand, they both analyzed the solution to the initial problem and then asked themselves how they could now come to new problems or what they could specifically vary about the game in order to come to new problems. Afterwards, they pose the variation that you may only remove 2 or 3 stones from the table. The following excerpt takes 1m 28s and shows the creation of this problem.

As in the section before, the codes for the metacognitive behavior are already set and marked in color. The coding is commented on after the transcript.

1	W:	We can take 2 or 3 stones from it.	P1
2	V:	2 or 3?	
3	W:	Mhmm.	
4	V:	And how many are there? (points to the pile of stones)	
5	W:	(laughs) I have not counted. I just took it. (Wenke counts off 20 stones.) 20 are here already.	
6	V:	(counts down 12 stones.) 12.	
7	W:	(counts off 10 stones.) 32. 42.	
8	V:	Yes.	
9	W:	How many do we take? 30?	MC2
10	V:	Yes. (5 seconds) Do you want to be Player A this time?	
11	W:	Yeah! (both start playing.) It would be terrific if there was only one left now (laughs). Unfortunately I can't make a move then (laughs).	MC3
12	V:	(Giggles. Afterwards both continue playing).	
13	W:	That certainly does not work. (3 seconds) It will not work (giggles).	
14	V:	(giggles)	
15	W:	Yes, but now you have taken especially so that it works out, right? (laughs) I'm not stupid!	
16	V:	(laughs)	
17	W:	Yes, that doesn't make much sense with 2 or 3.	E1
18	V:	No.	
19	W:	Okay, new idea. Now you can think of something.	MC1

Wenke brings a new problem into the focus in turn 1. This set of a new focus is coded as *planning* (P1). After counting how many stones they have on the table, Wenke asks in turn 9 how many stones to use for the new game. This statement is interpreted as controlling the representation of the posed problem, as the focus is already set in the form of a new game and they now have to agree on a number of stones at the beginning in order to be able to play the game (MC2). Wenke notices in turn 11 that the change they have posed can cause a situation in which one does not know who will win. This assessment of the consequences of the new game through their modification is coded as *monitoring & control* (MC3). In turn 17, Wenke says that their new problem does not make sense because of this new situation in which nobody wins. This statement is coded as *evaluation* since Wenke doesn't see much meaning in the new game where an undecidable situation can occur (E1). Wenke obviously does not intend to pursue this problem further in turn 19 and suggests that a new problem should be posed. She has a steering effect on the process which is why this statement is coded as *monitoring & control* (MC1).

Overall, in this process, Valerie and Wenke show noticeably insecurities. This can be seen through the numerous occurrences of laughter. Nevertheless, they also show metacognitive behavior – although less frequently than Tino and Ulrich. Like Tino and Ulrich, they realize during the game that an inevitable situation could occur. In Turn 17, Wenke also evaluates this, but they see this situation as a reason to reject the posed game.

9.6 Discussion

Research objective (1): Development of a framework for identifying problem-posing-specific aspects of metacognitive behavior (i.e. *planning, monitoring & control*, and *evaluating*) in pre-service teachers' problem-posing processes

Tables 9.3, 9.4, and 9.5 summarize the identified metacognitive activities. In total, four *planning* activities, four *monitoring & control* activities, and two *evaluating* activities were identified. Some of these activities may be considered as cognitive, but being able to intentionally use these kinds of cognitive behavior is a sign for metacognitive abilities. However, when metacognitive behavior is mentioned here, it always means the primarily metacognitive behavior in interaction with cognitive behavior. For example, searching for a solution can be seen as cognitive behavior, but considering the solution in order to get a better idea whether the posed problem is, for example, solvable or appropriate for a specific target group can be seen as metacognitive behavior. Most of the identified activities are indeed problem-posing specific. However, there are also activities (e.g., P3: Reflect necessary knowledge) that are not problem-posing specific. Moreover, not all codes (i.e., subcategories) within the superordinate categories of *planning*, *monitoring & control*, and *evaluating* are separable from each other. However, a clear separation between the superordinate categories should be recognizable.

We want to highlight parallels and differences between the category system for classifying teacher and students metacognitive activities in class discussions by Cohors-Fresenborg and Kaune (2007) (see Table 9.1) and our developed framework. The category P1 (*indication of a focus of attention, in particular with regard to tools/methods to be used or (intermediate) results or representations to be achieved*) in their system can be found in the categories P1 and P2 (see Table 9.3) in the problem-posing-specific framework developed here. P2 is considered separately as it is a central component of problem posing (Brown & Walter, 2005; Baumanns & Rott, 2022b). P4 (*express general procedure for problem posing*) corresponds to a specification for problem posing of the category P2 in table 9.1. For *monitoring & control*, MC1 represents a specification for problem posing of the category M8 and MC4 represents a specification for problem posing of the category M1. Categories M2 and M3 were merges into the problem-posing-specific category M2 and categories M4–M7 were merges into the problem-posing-specific category M3. Finally, for *evaluation*, rather rough parallels can be drawn between categories R1–R4 and E1 as well as R5–R7 and E2.

Research objective (2): Application of the framework developed in (1) as a proof of concept to make differences regarding metacognitive behavior in problem-posing processes visible

The analysis of metacognitive behavior in the problem-posing processes of Tino and Ulrich as well as Valerie and Wenke revealed differences. Tino and Ulrich's process had a greater frequency and density of metacognitive behavior compared to Valerie and Wenke's process. However, the coding does not allow for a statement about the depth of the metacognitive behavior, that is how sophisticated the specific metacognitive behaviors are. For example, in both processes the posed problem was *evaluated* (E1 and/or E2). In the process by Valerie and Wenke, the evaluation remains a single mention of futility in Turn 17 from which they discard the problem. Tino and Ulrich take their thoughts further and relate their evaluation to the new possible outcome of the game, which differs from the initial task (Turn 13, 17, 18). They even reflect on whether they want to restore similarity to the initial task by modifying it further (Turns 20, 21, 26, and 27). In contrast to Valerie and Wenke, they do not discard the problem. From a quantitative perspective, the observed processes do not claim to be representative. Therefore, a quantitative counting on the frequency of the individually occurring metacognitive behaviors would yield only insufficiently helpful new insights.

The perspective on the metacognitive behavior of both pairs reveals differences between the processes. Since in the selected excerpts the posed problems are rather identical, no differences would have been attested by looking only at the products. It also matters that Tino and Ulrich have a better understanding of the posed problem by suggesting a solution strategy as well as their motivational beliefs. However, the task did not explicitly ask for solving the posed problems. Tino's and Ulrich's impetus to do so independently may attests motivational beliefs that are certainly relevant in the context of metacognitive behavior.

9.7 Conclusion

The aim of the present explorative study was to investigate metacognitive behavior in problem-posing processes, which has been widely disregarded in problem-posing research. Analyses of 32 problem-posing processes of pre-service teacher students were conducted to identify metacognitive behaviors of *planning*, *monitoring & control*, and *evaluating*. Tables 9.3, 9.4, and 9.5 summarize these inductively developed categories of problem-posing-specific aspects of metacognitive behavior. In addition, two transcript excerpts were analyzed using the previously developed codes as a proof of concept to make differences regarding metacognitive behavior in problem-posing processes visible. Although, in both transcript excerpts, the product in terms of the posed problem is almost identical, the metacognitive behavior, as the analysis has shown, differs. The identified problem-posing-specific aspects of metacognitive behaviors enabled the disclosure of these differences. In addition, the analyses have shown that the consideration of metacognitive behavior allows a tentative assessment of the quality of the activity in general. This assessment is a new perspective on problem-posing processes.

Limitations of this study lie especially in the method of analyzing statements of pairs of students in a video-based content analysis to assess metacognitive behavior. As Goos et al. (2002) pointed out "student-student interactions could either help or hinder metacognitive decision making during paired problem solving, depending on students' flexibility in sharing metacognitive roles" (p. 197). This approach can be enriched in future studies by using stimulated recall interviews. Eye-tracking can serve as a potential stimulus of such interviews. Individual interviews with Thinking-Aloud approach was not used because Thinking Aloud, unpracticed, can interfere with the natural flow of such processes and, in particular, affects metacognitive activities (McKeown & Gentilucci, 2007), which could distort the analyses.

The framework developed in this study provides numerous opportunities for follow-up research on problem posing. As this study is based on problem-posing processes of student teachers, it would be of interest to see if additional problem-posing-specific aspects of metacognitive behaviors can be identified in a sample of, for example, pupils or expert problem posers. A larger analysis could also address the question of which metacognitive behaviors of planning, monitoring & control, and evaluating are particularly prevalent. As in research on problem solving, a comparison between metacognitive behaviors of experts and novices could reveal whether metacognitive behavior related to successful problem posing or whether there are substantial differences in metacognitive behavior in general. The data of this study was collected through structured problem-posing situations. Future studies could address the question of whether there are different metacognitive behaviors in unstructured situations or whether there are significant differences in the frequency of the different metacognitive behaviors between structured and unstructured problem-posing situations. Often, the ability to pose problems is measured by analyzing the products of a problem-posing process (Van Harpen & Sriraman, 2013; Bonotto, 2013; Singer et al., 2017). The analysis of metacognitive behavior could be used to assess the quality of problem posing on a process-oriented level. It is likely that there is a strong correlation between the products and the processes of problem posing. However, there may also be high quality processes by means of metacognitive behavior, but whose products attest lower quality because few and non-original problems were posed. Neglected in this study was the metacognitive facet knowledge about cognition. The importance of this facet of metacognition could also be the focus of future studies. Furthermore, metacognitive processes should be considered in addition to motivational beliefs (Zimmerman & Moylan, 2009). Those beliefs could certainly enrich the comparison of the two processes. Tino and Ulrich exhibit numerous behaviors indicative of their motivational beliefs. Even before the analyzed extract, Tino and Ulrich have numerous ideas about the Nim game. In turn 5 Ulrich shows that he would like to deal with the different ideas one after the other, presumably in order to do sufficient justice to all of them. Also, their numerous evaluative statements regarding their ideas (13; 17, 18, 25) as well as thinking their ideas further (20; 21; 26; 27) is an expression of intrinsic interest in posing new problems. In Valerie's and Wenke's process, such behaviors and especially the positive reference to the activity of problem posing are largely absent. Their laughter rather speaks for a general insecurity and a tendency towards low interest in problem posing. Future studies could focus more on this interplay between metacognitive behavior and motivational beliefs.

For teaching in school and university settings, these study's findings can be used to plan problem-posing activities in the classroom (cf. Kontorovich et al., 2012). For example, students could be encouraged to engage in metacognitive behavior during problem posing through appropriate construction of the problem-posing situation by the teacher. The teacher could ask the students to solve their posed problems or they could encourage reflection on the tasks by requiring students to pose tasks of varying difficulty for their classmates.

This study provides a first, qualitative insight into metacognitive behavior in problem posing. We hope that the perspective of metacognition will stimulate further studies in the field of problem posing research to gain further insights.

Part IV

Discussion & Outlook

Advanced organizer

In conclusion, two things should be accomplished in this discussion.

1. **Summary:** Within each article, the discussion and conclusion summarized the results of the individual studies. Within this discussion of the entire dissertation, the findings of all four journal articles will be summarized globally. We want to refer to the objectives of this PhD project as stated in chapter 2 on page 7.

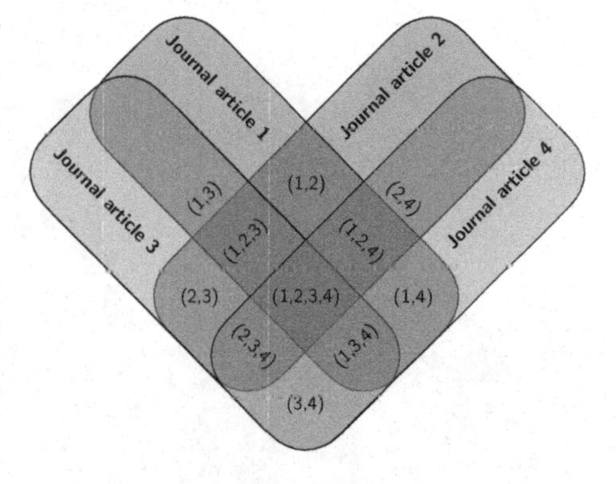

Fig. 9.1.: Venn diagram showing all possible intersections between the four journal articles

2. **Potential for further research:** Each article also highlighted the potential for future research based on the findings of that specific article. We also want to provide a more global outlook for this aspect in this chapter. For the different connections between the journal articles (see Figure 9.1), the potential for future research will be highlighted. In figure 9.1 and also in chapter 11 on page 233, these possible overlaps are indicated in the form of tuples. (1,2,3), for example, represents elaborations on the potential for future research with regard to the findings of journal articles 1, 2, and 3.

10 Summary

In the following, the findings of the individual journal articles based on the objectives set out in chapter 2 on page 7 are summarized.

(1) Conceptual part:
Systematic review of problem-posing situations and problem-posing activities.

Journal article 1:
In the first journal article, we were able to use a framework by Yeo (2017) to characterize 271 systematically collected problem-posing situations from empirical studies in terms of certain criteria of task openness. We have analyzed whether or not the problem-posing situations have a defined answer and whether or not they offer possibilities of extension. This enabled us to identify characteristic differences between problem-posing situations in terms of their openness and develop a concept of *mathematical problem posing.* In addition, we have found that Stoyanova and Ellerton's (1996) established categories of *free* and *semi-structured* for problem-posing situations can only be coded with less agreement. We therefore proposed to interpret *free* and *semi-structured* problem-posing situations on a spectrum of *unstructured* situations. At last, we found that the structured problem-posing situations differ in terms of their initial tasks. Some of them have routine tasks, some have problems as initial tasks.

L. Baumanns, *Mathematical Problem Posing*, Kölner Beiträge zur
Didaktik der Mathematik, https://doi.org/10.1007/978-3-658-39917-7_10

Journal article 2:

The second journal article was devoted to developing a framework that can be used to characterize problem-posing activities reported in empirical studies. We used three dimensions for this characterization: (1) problem posing as an activity of generating new or reformulating given problems, (2) emerging tasks on the spectrum between routine and non-routine problems, and (3) metacognitive behavior in problem-posing processes. For example, within the systematic review, we found that little low-level metacognitive activity within the problem-posing process has been reported in empirical studies.

(2) Empirical part:

Macroscopic and microscopic analysis of 32 problem-posing processes of pre-service mathematics teachers.

Journal article 3:

Journal article 3 bridges the conceptual part into the empirical part. The aim of the study was to develop a phase model for the macroscopic description of problem-posing processes. We want to identify recurring and distinguishable activities of 32 pairs of preservice teachers dealing with problem-posing situations. We have found that problem-posing processes can be described with five different phases: *situation analysis*, *variation*, *generation*, *problem solving*, and *evaluation*. The distinction between the episode types *variation* and *generation* has already been elaborated in dimension (1) of journal article 2. The episode type *evaluation* contains aspects of dimension (3) of journal article 2.

Journal article 4:

Journal article 4 deals with self-regulatory or metacognitive behavior within problem-posing processes. The result of this study is, at first, the identification of problem-posing specific metacognitive aspects of *planning*, *monitoring and control*, and *evaluation*. In addition, we used a case study to illustrate the extent to which these identified metacognitive behaviors reveal differences between problem-posing processes. Thus, this study takes up dimension (3) of journal article 2 as well as the episode type *evaluation* from journal article 3 in more depth.

11 Potential for further research

This chapter specifies future research areas that emerge from the findings, across the four journal articles. The subheadings are given as tuples such as (1,2,3) and denote future research based on, for example, the findings of journal articles 1, 2, and 3 (see figure 9.1 on page 227). Especially links between the conceptual and the empirical part of the present work have to be emphasized in this context.

(1,2): In article 1, problem-posing situations were analyzed in terms of how open they are, whether they are structured or unstructured, and whether a non-routine problem or a routine problem is given. In article 2, a three-dimensional framework for analyzing problem-posing activities reported on in empirical studies on problem posing was developed. This model combines three dimensions: (1) problem posing as an activity of generating new or reformulating given problems, (2) emerging tasks on the spectrum between routine and non-routine problems, and (3) metacognitive behavior in problem-posing processes. In conjunction with the findings in Journal Article 1, one might ask the following questions: Can the three-dimensional framework from article 2 be used in non-posing situations? If so, what areas of this framework are addressed the most? What areas of the three-dimensional framework are addressed in empirical studies when only unstructured situations are considered? Which areas of the three-dimensional framework are addressed in empirical studies when only structured situations with routine tasks or non-routine problems are considered?

L. Baumanns, *Mathematical Problem Posing*, Kölner Beiträge zur
Didaktik der Mathematik, https://doi.org/10.1007/978-3-658-39917-7_11

(1,3): Article 3 was devoted to the development of a descriptive phase model based on open and structured problem-posing situations with given non-routine problems. In future research projects, the findings from articles 1 and 3 could be used to investigate whether the phase model can also be applied to other situations. For example, can the phase model be applied at all if the non-posing situations identified in article 1 are used as initiators? Do the descriptions of the episode types of the phase model change if unstructured instead of structured problem-posing situations are used? Which differences can be seen in the processes when a routine problem is used instead of a non-routine problem in a structured problem-posing situation? First thoughts on some of these questions have already been concretized by students in the context of their master theses. However, intensive research still has to follow.

(1,4): Article 4 highlighted problem-posing-specific aspects of metacognitive behavior and its potential for identifying process-related differences. Combined with the findings from article 1, this raises similar questions to (1,3): Which problem-posing-specific aspects of metacognitive behavior can be found in non-posing situations? Which problem-posing-specific aspects of metacognitive behavior can be found in processes based on unstructured problem-posing situations? Which ones can be found when the initial task of structured situations is not a non-routine problem but a routine problem?

(2,3): The framework in article 2 has been applied only to information that can be obtained from articles of the respective studies. Dimension (1) from the framework of article 2 is explored in more depth in article 3. Perhaps the insights from article 3 regarding distinguishing between variation and generation can help in applying the framework from article 2 in empirical research.

(2,4): Dimensions (3) from the framework of article 2 are explored in more depth in article 4 of the present project. As in (2,3), it might be investigated: Perhaps the insights from article 4 regarding the problem-posing specific aspects of metacognitive behavior can help in applying the framework from article 2 in empirical research.

(3,4): Future research could also address the link between the macroscopic view from article 3 and the microscopic analysis from article 4. Schoenfeld (1985) has done something similar with problem solving. In his analyses, he found that successful problem-solving processes mostly involve self-regulatory – that is, metacognitive – behaviors. Do similar statements possibly hold when combining the findings from articles 3 and 4? At what points in the process do metacognitive activities emerge? Is this possibly related to the quality of the resulting products?

Furthermore, and following the work of Nowińska (2016), the macroscopic and microscopic analyses of the problem-posing processes could be the basis for assessing the quality of the problem-posing activity. Nowińska (2016) developed a two-stage evaluation system for assessing metacognitive discursive instructional quality in the classroom. Stage one uses the category system for classifying teacher and students metacognitive activities in class discussions by Cohors-Fresenborg and Kaune (2007) presented in journal article 3 (see table 9.1 on page 196). Stage two uses guiding questions based on seven high inference rating scales to evaluate metacognitive-discursive instructional quality. A similar approach could be used to evaluate the quality of problem-posing processes. Assessing the quality of the problem-posing process is still a recent topic in problem-posing research (Kontorovich & Koichu, 2016; Patáková, 2014; Rosli et al., 2013; Singer et al., 2017). A process-oriented approach on the future development of such guiding questions on the quality of problem-posing processes after analyzing these processes with the frameworks from journal articles 3 and 4 could help in exploring the quality of problem posing itself.

Part V

Listings

Bibliography

Abu-Elwan, R. (2002). Effectiveness of problem posing strategies on prospective mathematics teachers' problem solving performance. *Journal of Science and Mathematics Education in Southeast Asia*, *25*(1), 56–69. (Cited on p. 75.)

Arıkan, E. E. (2019). Comparison of Mathematical Problem Posing Skills of University Students in the Context of Critical Thinking Dispositions. *Istanbul Sabahattin Zaim University Journal of Faculty of Education*, *1*(1), 51–66. (Cited on pp. 120, 148.)

Arıkan, E. E., & Ünal, H. (2015). Investigation of Problem-Solving and Problem-Posing Abilities of Seventh-Grade Students. *Educational Sciences – Theory & Practice*, *15*(5), 1403–1416. (Cited on pp. 75, 76, 84, 89–91, 95, 119, 165.)

Artzt, A. F., & Armour-Thomas, E. (1992). Development of a Cognitive-Metacognitive Framework for Protocol Analysis of Mathematical Problem Solving in Small Groups. *Cognition and Instruction*, *9*(2), 137–175. (Cited on pp. 73, 137, 165, 168, 191, 197.)

Artzt, A. F., & Armour-Thomas, E. (1997). Mathematical problem solving in small groups: Exploring the interplay of students' metacognitive behaviors, perceptions, and ability levels. *Journal of Mathematical Behavior*, *16*(1), 63–74. (Cited on p. 202.)

Baumanns, L. (2023, accepted). Four mathematical miniatures on problem posing. In D. Sarikaya, K. Heuer, L. Baumanns, & B. Rott (Eds.), *Problem Posing and Solving for Mathematically Gifted and Interested Students – Best Practices, Research and Enrichment.* Springer.

L. Baumanns, *Mathematical Problem Posing*, Kölner Beiträge zur Didaktik der Mathematik, https://doi.org/10.1007/978-3-658-39917-7

Baumanns, L., & Rott, B. (2019). Is problem posing about posing „problems“? A terminological framework for researching problem posing and problem solving. In I. Gebel, A. Kuzle, & B. Rott (Eds.), *Proceedings of the 2018 Joint Conference of ProMath and the GDM Working Group on Problem Solving* (pp. 21–32). WTM. (Cited on pp. 95, 120, 144, 162.)

Baumanns, L., & Rott, B. (2020). Zum Prozess des Aufwerfens mathematischer Probleme – Validierung eines deskriptiven Prozessmodells. In L. Baumanns, J. Dick, A.-C. Söhling, N. Sturm, & B. Rott (Eds.), *Wat jitt dat, wenn et fädich es? Tagungsband der Herbsttagung des GDM-Arbeitskreises Problemlösen in Köln 2019* (pp. 87–100). WTM. (Cited on pp. 134, 137.)

Baumanns, L., & Rott, B. (2021). Rethinking problem-posing situations: a review. *Investigations in Mathematics Learning, 13*(2), 59–76. (Cited on pp. 74, 129, 162, 164, 166, 193.)

Baumanns, L., & Rott, B. (2022a). Developing a framework for characterizing problem-posing activities: a review. *Research in Mathematics Education, 24*(1), 28–50. (Cited on pp. 166, 191.)

Baumanns, L., & Rott, B. (2022b). The process of problem posing: development of a descriptive phase model of problem posing. *Educational Studies in Mathematics, 110*, 251–269. (Cited on pp. 198, 199, 203, 208, 219.)

Becker, J. P., & Shimada, S. (1997). *The open-ended approach: A new proposal for teaching mathematics.* National Council of Teachers of Mathematics. (Cited on p. 96.)

Behrens, R. (2018). *Formulieren und Variieren mathematischer Fragestellungen mittels digitaler Werkzeuge.* Springer. (Cited on p. 53.)

Berendonk, S. (2019). A Napoleonic Theorem for Trapezoids. *The American Mathematical Monthly, 126*(4), 367–369. (Cited on pp. 27, 33.)

Bicer, A., Lee, Y., Perihan, C., Capraro, M. M., & Capraro, R. M. (2020). Considering mathematical creative self-efficacy with problem posing as a measure of mathematical creativity. *Educational Studies in Mathematics, 105*(3), 457–485. (Cited on pp. 5, 82, 139, 164.)

Blum, W., & Leiß, D. (2007). How do students and teachers deal with modelling problems? In C. Haines, W. Blum, & P. Galbraith (Eds.), *Mathematical modelling (ICTMA 12): Education, engineering and economics* (pp. 222–231). Horwood. (Cited on pp. 68, 69.)

Bonotto, C. (2010). Realistic mathematical modeling and problem posing. In R. Lesh, P. Galbraith, C. Haines, & A. Hurford (Eds.), *Modeling Students' Mathematical Modeling Competencies* (pp. 399–408). Springer. (Cited on p. 68.)

Bonotto, C. (2013). Artifacts as sources for problem-posing activities. *Educational Studies in Mathematics, 83*(1), 37–55. (Cited on pp. 5, 53, 83, 161, 190, 222.)

Bonotto, C., & Santo, L. D. (2015). On the Relationship Between Problem Posing, Problem Solving, and Creativity in the Primary School. In F. M. Singer, N. F. Ellerton, & J. Cai (Eds.), *Mathematical Problem Posing. From Research to Effective Practice* (pp. 103–123). Springer. (Cited on pp. 5, 53, 75, 82.)

Bowen, G. A. (2009). Document Analysis as a Qualitative Research Method. *Qualitative Research Journal, 9*(2), 27–40. (Cited on p. 101.)

Brown, S. I., & Walter, M. I. (1983). *The Art of Problem Posing.* Franklin Institute Press. (Cited on pp. 3, 38, 39, 88, 89, 160, 192.)

Brown, S. I., & Walter, M. I. (2005). *The Art of Problem Posing* (3rd ed.). Lawrence Erlbaum Associates. (Cited on pp. 39, 41, 50, 124, 133, 134, 177, 206, 219.)

Bruder, R. (2000). Akzentuierte Aufgaben und heuristische Erfahrungen – Wege zu einem anspruchsvollen Mathematikunterricht für alle. In L. Flade & W. Herget (Eds.), *Mathematik. Lehren und Lernen nach TIMSS. Anregungen für die Sekundarstufen* (pp. 69–78). Volk und Wissen. (Cited on pp. 94, 108, 131.)

Butts, T. (1980). Posing Problems Properly. In S. Krulik & R. E. Reys (Eds.), *Problem solving in school mathematics* (pp. 23–33). NCTM. (Cited on pp. 38, 160, 192.)

Cai, J. (1998). An Investigation of U.S. and Chinese Students' Mathematical Problem Posing and Problem Solving. *Mathematics Education Research Journal, 10*(1), 37–50. (Cited on pp. 45, 75.)

Cai, J., & Hwang, S. (2002). Generalized and generative thinking in US and Chinese students' mathematical problem solving and problem posing. *Journal of Mathematical Behavior, 21*(4), 401–421. (Cited on pp. 4, 52, 75, 89–91, 94, 119.)

Cai, J., & Hwang, S. (2020). Learning to teach through mathematical problem posing: Theoretical considerations, methodology, and directions for future research. *International Journal of Educational Research, 102*, 1–8. (Cited on pp. 43, 46–48, 75, 95, 192.)

Cai, J., Hwang, S., Jiang, C., & Silber, S. (2015). Problem-posing Research in Mathematics Education: Some Answered and Unanswered Questions. In F. M. Singer, N. F. Ellerton, & J. Cai (Eds.), *Mathematical Problem Posing. From Research to Effective Practice* (pp. 3–34). Springer. (Cited on pp. 6, 75, 112, 125, 166, 192.)

Cai, J., & Jiang, C. (2017). An analysis of problem-posing tasks in Chinese and US elementary mathematics textbooks. *International Journal of Science and Mathematics Education, 15*(8), 1521–1540. (Cited on pp. 98, 108, 109.)

Cai, J., & Leikin, R. (2020). Affect in mathematical problem posing: conceptualization, advances, and future directions for research. *Educational Studies in Mathematics*, *105*(3), 287–301. (Cited on pp. 124, 139, 161, 164, 192.)

Cai, J., & Mamlok-Naaman, R. (2020). Posing researchable questions in mathematics and science education: purposefully questioning the questions for investigation. *International Journal of Science and Mathematics Education*, *18*(Supplement 1), S1–S7. (Cited on p. 41.)

Cai, J., Moyer, J. C., Wang, N., Hwang, S., Nie, B., & Garber, T. (2013). Mathematical problem posing as a measure of curricular effect on students' learning. *Educational Studies in Mathematics*, *83*(1), 57–69. (Cited on pp. 129, 130.)

Cambridge Dictionary. (2020). *Pose*. Retrieved July 1, 2020, from https://dictionary.cambridge.org/de/worterbuch/englisch/pose. (Cited on p. 93.)

Cankoy, O. (2014). Interlocked problem posing and children's problem posing performance in free structures situations. *International Journal of Science and Mathematics Education*, *12*(1), 219–238. (Cited on p. 142.)

Cantor, G. (1867). *De aequationibus secundi gradus indeterminatis*. Schultz. (Cited on pp. 37, 124, 160.)

Cantor, G. (1966). *Gesammelte Abhandlungen mathematischen und philosophischen Inhalts. Mit erläuternden Anmerkungen sowie mit Ergänzungen aus dem Briefwechsel Cantor-Dedekind* (Reprografischer Nachdruck der Ausgabe Berlin 1932). Georg Olms. (Cited on p. 88.)

Carrillo, J., & Cruz, J. (2016). problem-posing and questioning: two tools to help solve problems. In P. Felmer, E. Pehkonen, & J. Kilpatrick (Eds.), *Posing and Solving Mathematical Problems. Advances and New Perspectives* (pp. 24–36). Springer. (Cited on p. 199.)

Chen, L., Dooren, W. V., & Verschaffel, L. (2015). Enhancing the development of Chinese fifth-graders' problem-posing and problem-solving abilities, beliefs, and attitudes: a design experiment. In F. M. Singer, N. F. Ellerton, & J. Cai (Eds.), *Mathematical Problem Posing. From Research to Effective Practice* (pp. 309–329). Springer. (Cited on p. 75.)

Chen, L., Van Dooren, W., Chen, Q., & Verschaffel, L. (2011). An investigation on Chinese teachers' realistic problem posing and problem solving ability and beliefs. *International Journal of Science and Mathematics Education*, *9*(4), 919–948. (Cited on pp. 65, 75.)

Christou, C., Mousoulides, N., Pittalis, M., & Pitta-Pantazi, D. (2005). Problem solving and problem posing in a dynamic geometry environment. *The Mathematics Enthusiast*, *2*(2), 125–143. (Cited on p. 75.)

Christou, C., Mousoulides, N., Pittalis, M., Pitta-Pantazi, D., & Sriraman, B. (2005). An empirical taxonomy of problem posing processes. *ZDM - Mathematics Education*, *37*(3), 149–158. (Cited on pp. 98, 108, 109, 130, 146, 165, 198.)

Cifarelli, V. V., & Cai, J. (2005). The evolution of mathematical explorations in open-ended problem-solving situations. *Journal of Mathematical Behavior*, *24*(3-4), 302–324. (Cited on pp. 44, 60, 94, 96, 165.)

Cifarelli, V. V., & Sevim, V. (2015). Problem posing as reformulation and sense-making within probem solving. In F. M. Singer, N. F. Ellerton, & J. Cai (Eds.), *Mathematical Problem Posing. From Research to Effective Practice* (pp. 177–194). Springer. (Cited on pp. 57, 60, 77, 79, 80.)

Clement, J. (2000). Analysis of clinical interviews: Foundations and model viability. In A. Kelly & R. Lesh (Eds.), *Handbook of research design in mathematics and science education* (pp. 547–589). Lawrence Erlbaum Associates. (Cited on p. 170.)

Cohen, J. (1960). A coefficient of agreement for nominal scales. *Educational and Psychological Measurement, 20*(1), 37–46. (Cited on pp. 104, 172.)

Cohors-Fresenborg, E., & Kaune, C. (2007). Modelling classroom discussions and categorizing discursive and metacognitive activities. In D. Pitta-Pantazi & G. Philippou (Eds.), *Proceedings of the Fifth Congress of the European Society for Research in Mathematics Education* (pp. 1180–1189). University of Cyprus; ERME. (Cited on pp. 194–196, 203, 205, 218, 235.)

Cohors-Fresenborg, E., Kramer, S., Pundsack, F., Sjuts, J., & Sommer, N. (2010). The role of metacognitive monitoring in explaining differences in mathematics achievement. *ZDM – Mathematics Education, 42*(2), 231–244. (Cited on p. 194.)

Crespo, S. (2003). Learning to pose mathematical problems: Exploring changes in preservice teachers' practices. *Educational Studies in Mathematics, 52*(3), 243–270. (Cited on pp. 142, 148, 199.)

Crespo, S., & Harper, F. K. (2020). Learning to pose collaborative mathematics problems with secondary prospective teachers. *International Journal of Educational Research, 102*, 101430. (Cited on p. 164.)

Cross, D. R., & Paris, S. G. (1988). Developmental and instructional analyses of children's metacognition and reading comprehension. *Journal of Educational Psychology, 80*(2), 131–142. (Cited on pp. 136, 193.)

Cruz, M. (2006). A Mathematical Problem–Formulating Strategy. *International Journal for Mathematics Teaching and Learning*, 79–90. (Cited on pp. 65, 161, 166–168, 172, 182, 187, 198.)

Csikszentmihalyi, M., & Getzels, J. W. (1971). Discovery-oriented behavior and the originality of creative products. *Journal of Personality and Social Psychology, 19*, 47–52. (Cited on p. 82.)

da Ponte, J. P., & Henriques, A. (2013). Problem posing based on investigation activities by university students. *Educational Studies in Mathematics*, *83*(1), 145–156. (Cited on pp. 44, 60, 161, 164.)

Davis, P. J. (1985). What do i know? A study of mathematical self-awareness. *The College Mathematics Journal*, *16*(1), 22–41. (Cited on p. 80.)

Desoete, A., Roeyers, H., & Buysse, A. (2001). Metacognition and mathematical problem solving in Grade 3. *Journal of Learning Disabilities*, *34*(5), 435–447. (Cited on pp. 137, 191, 198.)

de Villiers, M. (1990). The role and function of proof in mathematics. *Pythagoras*, *24*, 17–24. (Cited on p. 71.)

Dickman, B. (2014). Problem posing with the multiplication table. *Journal of Mathematics Education at Teachers College*, *5*(1), 47–50. (Cited on pp. 75, 79, 95.)

Díez, F., & Moriyón, R. (2004). Solving mathematical exercises that involve symbolic computations. *Computing in Science & Engineering*, *6*(1), 81–84. (Cited on p. 94.)

Döring, N., & Bortz, J. (2016). *Forschungsmethoden und Evaluation in den Sozial- und Humanwissenschaften* (5th ed.). Springer. (Cited on p. 170.)

Doyle, W. (1988). Work in mathematics classes: The context of students' thinking during instruction. *Educational Psychologist*, *23*(2), 167–180. (Cited on p. 93.)

Duncker, K. (1945). On problem-solving. *Psychological Monographs*, *58*(5), i–113. (Cited on pp. 57, 78, 79.)

Einstein, A., & Infeld, L. (1938). *The evolution of physics: The Growth of Ideas from Early Concepts to Relativity and Quant*. Cambridge University Press. (Cited on p. 36.)

Ellerton, N. F. (1986). Children's made-up mathematics problems – A new perspective on talented mathematicians. *Educational Studies in Mathematics*, *17*, 261–271. (Cited on p. 38.)

Ellerton, N. F., Singer, F. M., & Cai, J. (2015). Problem posing in mathematics: reflecting on the past, energizing the present, and foreshadowing the future. In F. M. Singer, N. F. Ellerton, & J. Cai (Eds.), *Mathematical Problem Posing. From Research to Effective Practice* (pp. 547–556). Springer. (Cited on p. 191.)

English, L. D. (1997). The development of fifth-grade children's problem-posing abilities. *Educational Studies in Mathematics, 34*(3), 183–217. (Cited on pp. 88, 89, 124, 141.)

English, L. D. (1998). Children's problem posing within formal and informal contexts. *Journal for Research in Mathematics Education, 29*(1), 83–106. (Cited on p. 4.)

English, L. D. (2020). Teaching and learning through mathematical problem posing: commentary. *International Journal of Educational Research, 102*, 101451. (Cited on p. 69.)

English, L. D., Fox, J. L., & Watters, J. J. (2005). Problem posing and solving with mathematical modeling. *Teaching Children Mathematics, 12*(3), 156–163. (Cited on p. 68.)

Erdoğan, F., & Gül, N. (2020). An investigation of mathematical problem posing skills of gifted students. *Pegem Eğitim ve Öğretim Dergisi, 10*(3), 655–696. (Cited on p. 84.)

Erkan, B., & Kar, T. (2022). Pre-service mathematics teachers' problem-formulation processes: Development of the revised active learning framework. *Journal of Mathematical Behavior, 65*, 100918. (Cited on p. 200.)

Felmer, P., Pehkonen, E., & Kilpatrick, J. (Eds.). (2016). *Posing and solving mathematical problems. Advances and new perspectives.* Springer. (Cited on pp. 75, 102, 166.)

Fernandez, M. L., Hadaway, N., & Wilson, J. W. (1994). Problem solving: managing it all. *The Mathematics Teacher, 87*(3), 195–199. (Cited on pp. 73, 77, 78, 169.)

Fischer, W. (1863). Ein geometrischer Satz. *Archiv der Mathematik und Physik, 40*, 460–462. (Cited on p. 29.)

Fisher, J. C., Ruoff, D., & Shilleto, J. (1981). Polygons and Polynomials. In C. Davis, B. Grünbaum, & F. A. Sherk (Eds.), *The Geometric Vein* (pp. 321–333). Springer. (Cited on p. 33.)

Flavell, J. H. (1979). Metacognition and cognitive monitoring. A new area of cognitive-developmental inquiry. *American Psychologist, 34*(10), 906–911. (Cited on pp. 136, 182, 190, 193.)

Fleiss, J. L., Levin, B., & Pail, M. C. (2003). *Statistical methods for rates and proportions* (3rd ed.). John Wiley & Sons. (Cited on p. 111.)

Flick, U. (2007). *Managing quality in qualitative research.* Sage. (Cited on p. 204.)

Freudenthal, H. (1991). *Revisiting mathematics education.* Kluwer. (Cited on pp. 4, 161, 164.)

Fukushima, T. (2021). The role of generating questions in mathematical modeling. *International Journal of Mathematical Education in Science and Technology.* (Cited on p. 70.)

Garofalo, J., & Lester, F. K. (1985). Metacognition, cognitive monitoring, and mathematical performance. *Journal for Research in Mathematics Education, 16*(3), 163–176. (Cited on pp. 191, 197.)

Getzels, J. W., & Csikszentmihalyi, M. (1975). From problem solving to problem finding. In I. A. Taylor & J. W. Getzels (Eds.), *Perspectives in Creativity* (pp. 90–116). Routledge. (Cited on p. 82.)

Ghasempour, A. Z., Baka, M. N., & Jahanshahloo, G. R. (2013). Mathematical problem posing and metacognition: a theoretical framework. *International Journal of Pedagogical Innovations, 1*(2), 63–68. (Cited on p. 199.)

Gleich, S. (2019). Konzeption einer Studie zum Einfluss von Mathematik auf kreative Fähigkeiten. In A. Frank, S. Krauss, & K. Binder (Eds.), *Beiträge zum Mathematikunterricht 2019* (pp. 261–264). WTM. (Cited on p. 82.)

Goos, M., Galbraith, P., & Renshaw, P. (2002). Socially mediated metacognition: Creating collaborative zomes of proximal development in small group problem solving. *Educational Studies in Mathematics, 49*, 192–223. (Cited on pp. 202, 221.)

Greer, B. (1992). Multiplication and division as models of situations. In D. Grouws (Ed.), *Handbook of research on mathematics teaching and learning. A Project of the National Council of Teachers of Mathematics* (pp. 276–295). Macmillan. (Cited on p. 69.)

Groetsch, C. W. (2001). Inverse problems: The other two-thirds of the story. *Quaestiones Mathematicae, 24*(Supplement 1), 89–94. (Cited on p. 94.)

Guilford, J. P. (1967). *The nature of human intelligence.* McGraw-Hill. (Cited on pp. 5, 83.)

Hadamard, J. (1945). *The Psychology of Invention in the Mathematical Field.* Dover Publications. (Cited on pp. 124, 160, 192.)

Halmos, P. R. (1980). The heart of mathematics. *The American Mathematical Monthly, 87*(7), 519–524. (Cited on pp. 88, 124, 160, 192.)

Hanna, G. (2000). Proof, explanation and exploration: An overview. *Educational Studies in Mathematics, 44*, 5–23. (Cited on p. 71.)

Hartmann, L.-M., Krawitz, J., & Schukajlow, S. (2021). Create your own problem! When given descriptions of real-world situations, do students pose and solve modelling problems? *ZDM – Mathematics Education, 53*, 919–935. (Cited on p. 69.)

Haylock, D. (1997). Recognising mathematical creativity in schoolchildren. *ZDM – Mathematics Education, 29*(3), 68–74. (Cited on pp. 82, 83.)

Headrick, L., Wiezel, A., Tarr, G., Zhang, X., Cullicott, C. E., Middleton, J. A., & Jansen, A. (2020). Engagement and affect patterns in high school mathematics classrooms that exhibit spontaneous problem posing: an exploratory framework and study. *Educational Studies in Mathematics, 105*(3), 435–456. (Cited on pp. 161, 164, 188.)

Hersh, R. (1993). Proving is convincing and explaining. *Educational Studies in Mathematics, 24*, 389–399. (Cited on p. 71.)

Heuvel-Panhuizen, M. v. d., Middleton, J. A., & Streefland, L. (1995). Student-Generated Problems: Easy and Difficult Problems on Percentage. *For the Learning of Mathematics, 15*(3), 21–27. (Cited on p. 45.)

Hilbert, D. (1900). Mathematische Probleme. Vortrag, gehalten auf dem internationalen Mathematiker-Kongreß zu Paris 1900. *Nachrichten von der Königl. Gesellschaft der Wissenschaften zu Göttingen. Mathematisch-Physikalische Klasse, 3*, 253–297. (Cited on p. 37.)

Holle, H., & Rein, R. (2015). EasyDIAg: A tool for easy determination of interrater agreement. *Behavior Research Methods, 47*(3), 837–847. (Cited on pp. 173, 184.)

Jay, S., & Perkins, N. (1997). Problem finding: The search for mechanism. In M. A. Runco (Ed.), *The creativity research handbook* (pp. 257–293). Hampton Press. (Cited on pp. 82, 139.)

Jiang, C., & Cai, J. (2014). Collective problem posing as an emergent phenomenon in middle school mathematics group discourse. In P. Liljedahl, C. Nicol, S. Oesterle, & D. Allan (Eds.), *Proceedings of the 38th Conference of the International Group for the Psychology of Mathematics Education and the 36th Conference of the North American Chapter of the Psychology of Mathematics Education (Vol. 3)* (pp. 393–400). PME. (Cited on p. 115.)

Johnson, D., & Johnson, R. (1999). *Learning together and alone: cooperative, competitive and individualistic learning.* Prentice Hall. (Cited on pp. 170, 202.)

Joklitschke, J., Baumanns, L., & Rott, B. (2019). The intersection of problem posing and creativity: a review. In M. Nolte (Ed.), *Including the Highly Gifted and Creative Students – Current Ideas and Future Directions. Proceedings of the 11th International Conference on Mathematical Creativity and Giftedness (MCG 11)* (pp. 59–67). WTM. (Cited on pp. 83, 192.)

Kar, T., Özdemir, E., İpek, A. S., & Albayrak, M. (2010). The relation between the problem posing and problem solving skills of prospective elementary mathematics teachers. *Procedia - Social and Behavioral Sciences*, *2*(2), 1577–1583. (Cited on p. 75.)

Kaune, C. (2006). Reflection and metacognition in mathematics education – tools for the improvement of teaching quality. *ZDM – Mathematics Education*, *38*(4), 350–360. (Cited on pp. 194, 205.)

Kilpatrick, J. (1987). Problem formulating: where do good problems come from? In A. H. Schoenfeld (Ed.), *Cognitive Science and Mathematics Education* (pp. 123–147). Lawrence Erlbaum Associates. (Cited on pp. 3, 38, 139, 160, 192.)

Kilpatrick, J. (2016). Reformulating: approaching mathematical problem solving as inquiry. In P. Felmer, E. Pehkonen, & J. Kilpatrick (Eds.), *Posing and Solving Mathematical Problems. Advances and New Perspectives* (pp. 69–81). Springer. (Cited on p. 78.)

Kim, Y. R., Park, M. S., Moore, T. J., & Varma, S. (2013). Multiple levels of metacognition and their elicitation through complex problem-solving tasks. *Journal of Mathematical Behavior*, *37*, 337–396. (Cited on pp. 137, 198.)

Kinda, S. (2013). Generating scenarios of division as sharing and grouping: a study of Japanese elementary and university students. *Mathematical Thinking and Learning*, *15*(3), 190–200. (Cited on p. 107.)

Kılıç, Ç. (2017). A new problem-posing approach based on problem-solving strategy: analyzing pre-service primary school teachers' performance. *Educational Sciences – Theory & Practice, 17*(3), 771–789. (Cited on p. 75.)

Klinshtern, M., Koichu, B., & Berman, A. (2015). What Do High School Teachers Mean by Saying "I Pose My Own Problems"? In F. M. Singer, N. F. Ellerton, & J. Cai (Eds.), *Mathematical Problem Posing. From Research to Effective Practice* (pp. 449–467). Springer. (Cited on pp. 43, 44, 48.)

Koichu, B. (2020). Problem posing in the context of teaching for advanced problem solving. *International Journal of Educational Research, 102*, 101428. (Cited on p. 44.)

Koichu, B., Berman, A., & Moore, M. (2007). Heuristic literacy development and its relation to mathematical achievements of middle school students. *Instructional Science, 35*, 99–139. (Cited on p. 94.)

Koichu, B., & Kontorovich, I. (2013). Dissecting success stories on mathematical problem posing: a case of the Billiard Task. *Educational Studies in Mathematics, 83*(1), 71–86. (Cited on pp. 53, 113, 150, 161, 164, 167, 168, 182, 198.)

Konrad, K. (2010). Lautes Denken. In G. Mey & K. Mruck (Eds.), *Handbuch Qualitative Forschung in der Psychologie* (pp. 476–490). VS Verlag für Sozialwissenschaften. (Cited on p. 170.)

Kontorovich, I. (2020). Problem-posing triggers or where do mathematics competition problems come from? *Educational Studies in Mathematics.* (Cited on p. 65.)

Kontorovich, I., & Koichu, B. (2012). Feeling of innovation in expert problem posing. *Nordic Studies in Mathematics Education, 17*(3–4), 199–212. (Cited on p. 5.)

Kontorovich, I., & Koichu, B. (2016). A case study of an expert problem poser for mathematics competitions. *International Journal of Science and Mathematics Education*, *14*(1), 81–99. (Cited on pp. 5, 65, 126, 127, 154, 163, 187, 235.)

Kontorovich, I., Koichu, B., Leikin, R., & Berman, A. (2012). An exploratory framework for handling the complexity of mathematical problem posing in small groups. *Journal of Mathematical Behavior*, *31*(1), 149–161. (Cited on pp. 5, 53, 76, 77, 95, 137, 148, 150, 165, 183, 191, 198–200, 208, 222.)

Kopparla, M., Bicer, A., Vela, K., Lee, Y., Bevan, D., Kwon, H., Caldwell, C., Capraro, M. M., & Capraro, R. M. (2018). The effects of problem-posing intervention types on elementary students' problem-solving. *Educational Studies*. (Cited on p. 75.)

Kuhn, D., & Dean, D. (2004). Metacognition: A bridge between cognitive psychology and educational practice. *Theory into Practice*, *43*(4), 268–273. (Cited on pp. 136, 193.)

Kuzle, A. (2013). Patterns of metacognitive behavior during mathematics problem-solving in a dynamic geometry environment. *International Electronic Journal of Mathematics Education*, *8*(1), 20–40. (Cited on pp. 191, 198.)

Kwek, M. L. (2015). Using problem posing as a formative assessment tool. In F. M. Singer, N. F. Ellerton, & J. Cai (Eds.), *Mathematical Problem Posing. From Research to Effective Practice* (pp. 273–292). Springer. (Cited on p. 5.)

Landis, J. R., & Koch, G. G. (1977). The measurement of observer agreement for categorical data. *Biometrics*, *33*(1), 159–174. (Cited on pp. 142, 185.)

Lang, S. (1989). *Faszination Mathematik – Ein Wissenschaftler stellt sich der Öffentlichkeit*. Vieweg. (Cited on pp. 37, 88, 124, 192.)

Leavy, A., & Hourigan, M. (2020). Posing mathematically worthwhile problems: developing the problem-posing skills of prospective teachers. *Journal of Mathematics Teacher Education*, *23*, 341–361. (Cited on p. 147.)

Lee, S.-Y. (2020). Research status of mathematical problem posing in mathematics education journals. *International Journal of Science and Mathematics Education*. (Cited on pp. 124, 192.)

Leikin, R. (2015). Problem posing for and through investigations in a dynamic geometry environment. In F. M. Singer, N. F. Ellerton, & J. Cai (Eds.), *Mathematical Problem Posing. From Research to Effective Practice* (pp. 373–391). Springer. (Cited on p. 72.)

Leikin, R., & Elgrably, H. (2020). Problem posing through investigations for the development and evaluation of proof-related skills and creativity skills of prospective high school mathematics teachers. *International Journal of Educational Research*. (Cited on pp. 44, 72, 82.)

Leikin, R., & Grossman, D. (2013). Teachers modify geometry problems: from proof to investigation. *Educational Studies in Mathematics*, *82*(3), 515–531. (Cited on p. 72.)

Leikin, R., & Lev, M. (2013). Mathematical creativity in generally gifted and mathematically excelling adolescents: What makes the difference? *ZDM - Mathematics Education*, *45*(2), 183–197. (Cited on pp. 5, 94.)

Lesh, R., & Zawojekski, J. (2007). Problem solving and modeling. In F. K. Lester (Ed.), *Second Handbook of Research on Mathematics Teaching and Learning* (2nd ed., pp. 763–804). NCTM. (Cited on p. 94.)

Lester, F. K. (1980). Research on mathematical problem solving. In R. J. Shumway (Ed.), *Research in mathematics education* (pp. 286–323). NCTM. (Cited on p. 73.)

Lester, F. K. (2005). On the theoretical, conceptual, and philosophical foundations for research in mathematics education. *ZDM – Mathematics Education*, *37*(6), 457–467. (Cited on p. 162.)

Lester, F. K., Garofalo, J., & Kroll, D. L. (1989). *The role of metacognition in mathematical problem solving: A study of two grade seven classes. Final Report.* Indiana University Mathematics Education Development Center. (Cited on p. 137.)

Leuders, T. (2015). Aufgaben in Forschung und Praxis. In R. Bruder, L. Hefendehl-Hebeker, B. Schmidt-Thieme, & H.-G. Weigand (Eds.), *Handbuch der Mathematikdidaktik* (pp. 435–460). Springer. (Cited on p. 94.)

Leung, S.-k. S. (1993a). Mathematical problem posing: The influence of task formats, mathematics knowledge, and creative thinking. In J. Hirabayashi, N. Nohda, K. Shigematsu, & F. Lin (Eds.), *Proceedings of the 17th International Conference of the International Group for the Psychology of Mathematics Education* (pp. 123–147). Lawrence Erlbaum Associates. (Cited on p. 5.)

Leung, S.-k. S. (1993b). *The relation of mathematical knowledge and creative thinking to the mathematical problem posing of prospective elementary school teachers on tasks differing in numerical information content.* Unveröffentlichte Dissertation. (Cited on p. 82.)

Leung, S.-k. S. (1994). On analyzing problem-posing processes. a study of prospective elementary teachers differing in mathematics knowledge. In J. P. da Ponte & J. F. Matos (Eds.), *Proceedings of the 18th International Conferrence of the International Group for the Psychology of Mathematics Education. Volume III* (pp. 168–175). PME. (Cited on p. 5.)

Leung, S.-k. S. (1997). On the role of creative thinking in problem posing. *ZDM – Mathematics Education*, *29*(3), 81–85. (Cited on pp. 4, 82, 83.)

Leung, S.-k. S. (2013). Teachers implementing mathematical problem posing in the classroom: challenges and strategies. *Educational Studies in Mathematics*, *83*(1), 103–116. (Cited on pp. 77, 79, 113.)

Liljedahl, P., & Cai, J. (2021). Empirical research on problem solving and problem posing: a look at the state of the art. *ZDM – Mathematics Education.* (Cited on p. 75.)

Lowrie, T. (2002). Young children posing problems: The influence of teacher intervention on the type of problems children pose. *Mathematics Education Research Journal*, *14*(2), 87–98. (Cited on p. 113.)

Martinez-Luaces, V., Fernandez-Plaza, J., Rico, L., & Ruiz-Hildalgo, J. F. (2019a). Inverse reformulations of a modelling problem proposed by prospective teachers in Spain. *International Journal of Mathematical Education in Science and Technology, online.* (Cited on pp. 45, 61, 151.)

Martinez-Luaces, V., Fernandez-Plaza, J., Rico, L., & Ruiz-Hildalgo, J. F. (2019b). Inverse reformulations of a modelling problem proposed by prospective teachers in Spain. *International Journal of Mathematical Education in Science and Technology, online.* (Cited on p. 61.)

Mason, J. (2000). Asking mathematical questions mathematically. *International Journal of Mathematical Education in Science and Technology*, *31*(1), 91–111. (Cited on p. 139.)

Mason, J., Burton, L., & Stacey, K. (1982). *Thinking Mathematically.* Pearson Prentice Hall. (Cited on p. 165.)

Mason, J., Burton, L., & Stacey, K. (2010). *Thinking Mathematically* (2nd ed.). Pearson. (Cited on pp. 73, 94.)

Mayring, P. (2000). Qualitative Content Analysis. *Forum: Qualitative Social Research*, *1*(2), Art. 20. (Cited on p. 203.)

Mayring, P. (2014). *Qualitative content analysis. Theoretical foundation, basic procedures and software solution.* (Cited on p. 172.)

McKeown, R. G., & Gentilucci, J. L. (2007). Think-aloud strategy: metacognitive development and monitoring comprehension in the middle school second-language classroom. *Journal of Adolescent & Adult Literacy, 51*(2), 136–147. (Cited on p. 221.)

Meschkowski, H. (1990). *Denkweisen großer Mathematiker. Ein Weg zur Geschichte der Mathematik.* Vieweg. (Cited on p. 70.)

Mevarech, Z., & Fridkin, S. (2006). The effects of IMPROVE on mathematical knowledge, mathematical reasoning and meta-cognition. *Metacognition and Learning, 1*(1), 85–97. (Cited on p. 197.)

Moher, D., Liberati, A., Tetzlaff, J., & Altman, D. G. (2009). Preferred reporting items for systematic reviews and meta-analyses: the prisma statement. *PLoS Medicine, 6*(7), e1000097. (Cited on pp. 101, 137.)

NCTM. (2000). *Principles and standards for school mathematics.* NCTM. (Cited on p. 124.)

Newell, A., & Simon, H. A. (1972). *Human Problem Solving.* Prentice-Hall. (Cited on p. 93.)

Nicol, C. C., & Crespo, S. (2006). Learning to teach with mathematics textbooks: How preservice teachers interpret and use curriculum materials. *Educational Studies in Mathematics, 62*(3), 331–355. (Cited on p. 65.)

Nishiyama, Y. (2011). The beautiful geometric theorem of van Aubel. *International Journal of Pure and Applied Mathematics, 66*(1), 71–80. (Cited on p. 34.)

Nowińska, E. (2016). The design of a high inference rating system for an evaluation of metacognitiv-discursive instructional quality. In S. Zehetmeier, B. Rösken-Winter, D. Potari, & M. Ribeiro (Eds.), *ERME Topic Conference on Mathematics Teaching, Resources and Teacher Professional Development* (pp. 46–55). (Cited on p. 235.)

Nuha, M. A., Waluya, S. B., & Junaedi, I. (2018). Mathematical creative process Wallas model in students problem posing with lesson study approach. *International Journal of Instruction, 11*(2), 527–538. (Cited on p. 82.)

Özsoy, G., & Ataman, A. (2009). The effect of metacognitive strategy training on mathematical problem solving achievement. *International Electronic Journal of Elementary Education, 1*(2), 67–82. (Cited on pp. 191, 198.)

Padberg, F. (2008). *Elementare Zahlentheorie* (3rd ed.). Springer. (Cited on p. 61.)

Papadopoulos, I., Patsiala, N., Baumanns, L., & Rott, B. (2022). Multiple approaches to problem posing: Theoretical considerations to its definition, conceptualization, and implementation. *CEPS Journal, 12*(1), 13–34. (Cited on pp. 68, 131.)

Patáková, E. (2014). Expert Recurrence of Linear Problem Posing. *Procedia – Social and Behavioral Sciences, 152*, 590–595. (Cited on pp. 5, 161, 164, 187, 235.)

Pehkonen, E. (1995). Introduction: Use of open-ended problems. *ZDM – Mathematics Education, 27*(2), 55–57. (Cited on pp. 76, 94, 96.)

Pelczer, I. (2008). Problem posing strategies of mathematically gifted students. In R. Leikin (Ed.), *Proceedings of the 5th International Conference on Creativity in Mathematics and the Education of Gifted Students* (pp. 193–199). CET. (Cited on p. 84.)

Pelczer, I., & Gamboa, F. (2009). Problem posing: Comparison between experts and novices. In M. Tzekaki, M. Kaldrimidou, & H. Sakonidis (Eds.), *Proceedings of the 33th Conference of the International Group for the Psychology of Mathematics Education. Vol. 4* (pp. 353–360). PME. (Cited on pp. 5, 79, 80, 95, 137, 161, 166–168, 172, 181, 182, 187, 191, 198–200, 206, 208.)

Pelczer, I., & Rodríguez, F. G. (2011). Creativity assessment in school settings through problem posing tasks. *The Mathematics Enthusiast*, *8*(1&2), 383–398. (Cited on pp. 82, 161, 164.)

Pintrich, P. R. (2000). The role of goal orientation in self-regulated learning. In M. Boekaerts, P. R. Pintrich, & M. Zeidner (Eds.), *Handbook of Self-Regulation* (pp. 451–529). Academic Press. (Cited on p. 194.)

Pintrich, P. R. (2002). The role of metacognitive knowledge in learning, teaching, and assessing. *Theory Into Practice*, *41*(4), 219–225. (Cited on pp. 136, 193.)

Poincaré, H. (1973). *Wissenschaft und Methode* (unveränderter reprografischer Nachdruck der Ausgabe von 1914). Wissenschaftliche Buchgesellschaft. (Cited on p. 37.)

Pólya, G. (1957). *How to solve it. A new aspect of mathematical method* (2nd ed.). University Press. (Cited on pp. 3, 23, 37, 38, 54, 56, 73, 74, 77–79, 88, 133, 148, 154, 165, 166, 168, 172, 179, 192, 194, 197.)

Pólya, G. (1966). On teaching problem solving. In The Conference Board of the Mathematical Sciences (Ed.), *The Role of Axiomatics and Problem Solving in Mathematics* (pp. 123–129). Ginn. (Cited on pp. 95, 105, 135, 136, 162.)

Posamentier, A. S., & Krulik, S. (2008). *Problem-solving strategies for efficient and elegant solutions, grades 6-12: a resource for the mathematics teacher* (2nd ed.). Corwin. (Cited on p. 94.)

Poulos, A. (2017). A research on the creation of problems for mathematical competitions. *The Teaching of Mathematics*, *20*(1), 26–36. (Cited on p. 65.)

Priest, D. J. (2009). *A problem-posing intervention in the development of problem-solving competence of underachieving, middle-year students* (Doctoral dissertation). Queensland University of Technology. Brisbane. (Cited on p. 75.)

Ramanujam, R. (2013). Paul Erdős. The artist of problem-posing. *At Right Angles*, *2*(2), 5–10. (Cited on p. 37.)

Renzulli, J. S. (2005). The three-ring conception of giftedness: A developmental model for promoting creative productivity. In R. J. Sternberg & J. E. Davidson (Eds.), *Conceptions of Giftedness* (2nd ed., pp. 246–279). Cambridge University Press. (Cited on pp. 83, 84.)

Renzulli, J. S., & Reis, S. M. (2021). The three ring conception of giftedness: A change in direction from being gifted to the development of gifted behaviors. In R. J. Sternberg & D. Ambrose (Eds.), *Conceptions of Giftedness and Talent* (pp. 335–357). Palgrave MacMillan. (Cited on p. 83.)

Rosli, R., Capraro, M. M., Goldsby, D., Gonzalez, E., Onwuegbuzie, A. J., & Capraro, R. M. (2015). Middle-grade preservice teachers‘ mathematical problem solving and problem posing. In F. M. Singer, N. F. Ellerton, & J. Cai (Eds.), *Mathematical Problem Posing. From Research to Effective Practice* (pp. 333–354). Springer. (Cited on p. 75.)

Rosli, R., Goldsby, D., & Capraro, M. M. (2013). Assessing students' mathematical problem-solving and problem-posing skills. *Asian Social Science*, *9*(16), 54–60. (Cited on pp. 75, 187, 235.)

Rott, B. (2012). Problem solving processes of fifth graders – an analysis of problem solving types. *Proceedings of the 12th ICME Conference. Seoul, Korea* (pp. 3011–3021). (Cited on p. 135.)

Rott, B. (2013). Process regulation in the problem-solving processes of fifth graders. *CEPS Journal*, *3*(4), 25–39. (Cited on p. 191.)

Rott, B. (2014). Mathematische Problembearbeitungsprozesse von Fünftklässlern – Entwicklung eines deskriptiven Phasenmodells. *Journal für Mathematik-Didaktik*, *35*(2), 251–282. (Cited on pp. 95, 114, 203.)

Rott, B., Specht, B., & Knipping, C. (2021). A descriptive phase model of problem-solving processes. *ZDM – Mathematics Education.* (Cited on pp. 73, 165, 168, 169, 187.)

Runco, M., & Nemiro, J. (1994). Problem finding, creativity, and giftedness. *Roeper Review*, *16*(4), 235–241. (Cited on pp. 82, 84.)

Ruthven, K. (2020). Problematising learning to teach through mathematical problem posing. *International Journal of Educational Research*, *102*, 1–7. (Cited on pp. 75, 110, 116, 131, 153, 191.)

Sayed, R. A.-E. E. (2002). Effectiveness of problem posing strategies on prospective mathematics teachers‘ problem solving performance. *Journal of Science and Mathematics Education in Southeast Asia*, *25*(1), 56–69. (Cited on p. 75.)

Schneider, W., & Artelt, C. (2010). Metacognition and mathematics education. *ZDM – Mathematics Education*, *42*(2), 149–161. (Cited on pp. 190, 197.)

Schoenfeld, A. H. (1985a). Making sense of "out loud" problem-solving protocols. *Journal of Mathematical Behavior*, *4*, 171–191. (Cited on pp. 170, 202.)

Schoenfeld, A. H. (1985b). *Mathematical Problem Solving.* Academic Press. (Cited on pp. 38, 41, 54, 56, 73, 74, 88, 92, 94, 132, 135, 154, 162, 165, 168, 169, 171, 172, 181, 182, 184, 187, 197.)

Schoenfeld, A. H. (1987). What's all the fuss about metacognition? In A. H. Schoenfeld (Ed.), *Cognitive Science and Mathematics Education* (pp. 189–215). Lawrence Erlbaum Associates. (Cited on pp. 137, 191, 197.)

Schoenfeld, A. H. (1989). Teaching mathematical thinking and problem solving. In L. B. Resnick & L. E. Klopfer (Eds.), *Toward a thinking curriculum: Current cognitive Research* (pp. 83–103). Association for Supervisors; Curriculum Developers. (Cited on pp. 38, 95, 135.)

Schoenfeld, A. H. (1992). Learning to think mathematically: Problem solving, metacognition and sense – making in mathematics. In D. Grouws (Ed.), *Handbook of research on mathematics teaching and learning* (pp. 334–370). MacMillan. (Cited on pp. 38, 73, 93, 134, 197.)

Schoenfeld, A. H. (2000). Purposes and methods of research in mathematics education. *Notices of the AMS*, *47*, 641–649. (Cited on pp. 6, 154, 162, 169, 185.)

Schoenfeld, A. H. (2020). On meaningful, researchable, and generative questions. *International Journal of Science and Mathematics Education*, *18*, 67–82. (Cited on p. 41.)

Schraw, G., & Moshman, D. (1995). Metacognitive theories. *Educational Psychology Review*, *7*(4), 351–371. (Cited on pp. 136, 137, 182, 193, 194, 203.)

Schreck, A., & Baumanns, L. (2022). Nicht verzagen, selbst was fragen! Digitale Hilfsmittel als adaptives Werkzeug. *mathematik lehren*, *233*, 23–27.

Schupp, H. (2002). *Thema mit Variationen. Aufgabenvariationen im Mathematikunterricht.* Franzbecker. (Cited on pp. 41, 48, 63, 171, 202.)

Shapira, Z. (2011). "I've got a theory paper—do you?": Conceptual, empirical, and theoretical contributions to knowledge in the organizational sciences. *Organization Science*, *22*(5), 1312–1321. (Cited on p. 7.)

Sharygin, I. F. (2001). The art of porsing novel problems. *Quantum*, *8*(2), 12–21. (Cited on p. 65.)

Silber, S., & Cai, J. (2017). Pre-service teachers' free and structured mathematical problem posing. *International Journal of Mathematical Education in Science and Technology, 48*(2), 163–184. (Cited on pp. 107, 108.)

Silver, E. A. (1994). On mathematical problem posing. *For the Learning of Mathematics, 14*(1), 19–28. (Cited on pp. 3, 4, 41, 43, 46–48, 77, 79, 82, 88, 92, 94, 109, 124, 128, 129, 131, 132, 160, 162, 172, 182, 187, 192.)

Silver, E. A. (1995). The nature and use of open problems in mathematics education: Mathematical and pedagogical perspectives. *ZDM – Mathematics Education, 27*(2), 67–72. (Cited on pp. 76, 77, 95, 96, 119, 165, 198.)

Silver, E. A. (1997). Fostering creativity through instruction rich in mathematical problem solving and problem posing. *ZDM – Mathematics Education, 29*(3), 75–80. (Cited on pp. 3, 5, 82, 160, 192.)

Silver, E. A. (2013). Problem-posing research in mathematics education: looking back, looking around, and looking ahead. *Educational Studies in Mathematics, 83*(1), 157–162. (Cited on pp. 4, 7, 125, 192.)

Silver, E. A., & Cai, J. (1996). An analysis of arithmetic problem posing by middle school students. *Journal for Research in Mathematics Education, 27*(5), 521–539. (Cited on pp. 5, 75.)

Silver, E. A., Mamona-Downs, J., Leung, S., & Kenney, P. (1996). Posing mathematical problems: An exploratory study. *Journal for Research in Mathematics Education, 27*(3), 293–309. (Cited on pp. 4, 53.)

Singer, F. M., Ellerton, N. F., & Cai, J. (2013). Problem-posing research in mathematics education: new questions and directions. *Educational Studies in Mathematics, 83*(1), 1–7. (Cited on pp. 7, 89, 125.)

Singer, F. M., Ellerton, N. F., & Cai, J. (Eds.). (2015). *Mathematical Problem Posing. From Research to Effective Practice*. Springer. (Cited on pp. 102, 166.)

Singer, F. M., Pelczer, I., & Voica, C. (2011). Problem posing and modification as a criterion of mathematical creativity. In M. Pytlak, T. Rowland, & E. Swoboda (Eds.), *Proceedings of the CERME 7* (pp. 1133–1142). University of Rzeszów. (Cited on p. 82.)

Singer, F. M., & Voica, C. (2015). Is problem posing a tool for identifying and developing mathematical creativity? In F. M. Singer, N. F. Ellerton, & J. Cai (Eds.), *Mathematical Problem Posing. From Research to Effective Practice* (pp. 141–174). Springer. (Cited on pp. 82, 160, 192, 199.)

Singer, F. M., & Voica, C. (2017). When mathematics meets real objects: How does creativity interact with expertise in problem solving and posing? *Creativity and Giftedness*, 75–103. (Cited on p. 82.)

Singer, F. M., Voica, C., & Pelczer, I. (2017). Cognitive styles in posing geometry problems: implications for assessment of mathematical creativity. *ZDM – Mathematics Education*, *49*(1), 37–52. (Cited on pp. 5, 82, 83, 146–148, 161, 187, 190, 222, 235.)

Smyth, M. R. F. (2007). MacCool's proof of napoleon's theorem. *Irish Mathematical Society Bulletin*, *59*, 71–77. (Cited on p. 29.)

Sriraman, B., & Dickman, B. (2017). Mathematical pathologies as pathways into creativity. *ZDM – Mathematics Education*, *49*(1), 137–145. (Cited on pp. 83, 140.)

Stein, M. K., Grover, B. W., & Henningsen, M. (1996). Building student capacity for mathematical thinking and reasoning: An analysis of mathematical tasks used in reform classrooms. *American Educational Research Journal*, *33*(2), 455–488. (Cited on p. 93.)

Stillman, G. (2015). Problem finding and problem posing for mathematical modelling. In N. H. Lee & D. K. E. Ng (Eds.), *Mathematical modelling. From theory to practice* (pp. 41–56). World Scientific. (Cited on p. 69.)

Stoyanova, E. (1997). *Extending and exploring students' problem solving via problem posing* (Doctoral dissertation). Edith Cowan University. (Cited on pp. 44, 47, 92, 97, 99, 112, 164, 171, 202.)

Stoyanova, E. (1999). Extending students' problem solving via problem posing. *Australian Mathematics Teacher*, *55*(3), 29–35. (Cited on pp. 97, 99, 193.)

Stoyanova, E., & Ellerton, N. F. (1996). A framework for research into students' problem posing in school mathematics. In P. C. Clarkson (Ed.), *Technology in mathematics education* (pp. 518–525). Mathematics Education Research Group of Australasia. (Cited on pp. 3, 43, 44, 46, 47, 52, 68, 92, 97, 98, 100, 104, 109, 110, 112, 120, 128, 129, 132, 133, 143, 146, 162, 164, 192, 193, 229.)

Sullivan, P., & Clarke, D. (1991). Catering to all abilities through 'good' questions. *Arithmetic Teacher*, *39*(2), 14–18. (Cited on pp. 116, 146, 147.)

Sweller, J., & Cooper, G. A. (1985). The use of worked examples as a substitute for problem solving in learning algebra. *Cognition and Instruction*, *2*, 59–89. (Cited on p. 94.)

Tao, T. (2006). *Solving mathematical problems: A personal perspective.* Oxford University Press. (Cited on p. 88.)

Tichá, M., & Hošpesová, A. (2013). Developing teachers' subject didactic competence through problem posing. *Educational Studies in Mathematics*, *83*(1), 133–143. (Cited on pp. 65, 111, 125, 126, 154, 199.)

Törner, G., & Arzarello, F. (2012). Grading mathematics education research journals. *EMS Newsletter*, (86), 52–54. (Cited on pp. 101, 137.)

Torrance, E. P. (1974). *Torrance Test of Creative Thinking.* Scholastic Testing Service. (Cited on pp. 5, 82, 83.)

Van der Stel, M., Veenman, M. V. J., Deelen, K., & Haenen, J. (2010). The increasing role of metacognitive skills in math: a cross-sectional study from a developmental perspective. *ZDM – Mathematics Education, 42*(2), 219–229. (Cited on pp. 194, 197.)

Van Gog, T., Paas, F., & Van Merriënboer, J. J. G. (2006). Effects of process-oriented worked examples on troubleshooting transfer performance. *Learning and Instruction, 16*(2), 154–164. (Cited on p. 94.)

Van Harpen, X., & Presmeg, N. C. (2013). An investigation of relationships between students' mathematical problem-posing abilities and their mathematical content knowledge. *Educational Studies in Mathematics, 83*(1), 117–132. (Cited on pp. 5, 83, 141, 148, 164.)

Van Harpen, X., & Sriraman, B. (2013). Creativity and mathematical problem posing: an analysis of high school students' mathematical problem posing in China and the USA. *Educational Studies in Mathematics, 82*(2), 201–221. (Cited on pp. 5, 82, 83, 160, 161, 190–192, 222.)

Vargyas, E.-T. (2020). Napoleonische Rechtecke. In E. Specht, E. Quaisser, & P. Bauermann (Eds.), *50 Jahre Bundeswettbewerb Mathematik. Die schönsten Aufgaben* (pp. 73–76). Springer. (Cited on p. 34.)

Veenman, M. V. J., Van Hout-Wolters, B. H. A. M., & Afflerbach, P. (2006). Metacognition and learning: conceptual and methodological considerations. *Metacognition Learning, 1*, 3–14. (Cited on p. 201.)

Voica, C., & Singer, F. M. (2013). Problem modification as a tool for detecting cognitive flexibility in school children. *ZDM – Mathematics Education, 45*(2), 267–279. (Cited on pp. 5, 60, 82, 83, 143.)

Voica, C., Singer, F. M., & Stan, E. (2020). How are motivation and self-efficacy interacting in problem-solving and problem-posing? *Educational Studies in Mathematics, 105*(3), 487–517. (Cited on pp. 137, 199.)

Wallas, G. (1926). *The art of thought.* Harcourt Brace. (Cited on pp. 81, 82.)

Walter, M. I., & Brown, S. I. (1977). Problem posing and problem solving: An illustration of their interdependence. *Mathematics Teacher, 70*(1), 4–13. (Cited on pp. 38, 75.)

Wessman-Enzinger, N. M., & Mooney, E. S. (2021). Conceptual models for integer addition and substraction. *International Journal of Mathematical Education in Science and Technology, 52*(3), 349–376. (Cited on p. 130.)

Whitebread, D., Coltman, P., Pasternak, D. P., Sangster, C., Grau, V., Bingham, S., Almeqdad, Q., & Demetriou, D. (2009). The development of two observational tools for assessing metacognition and self-regulated learning in young children. *Metacognition and Learning, 4*(1), 63–85. (Cited on pp. 136, 193, 203.)

Wickelgren, W. A. (1974). *How to Solve Problems.* W. H. Freeman; Company. (Cited on p. 93.)

Williams, S. R., & Leatham, K. R. (2017). Journal quality in mathematics education. *Journal for Research in Mathematics Education, 48*(4), 369–396. (Cited on pp. 101, 139.)

Wittmann, E. C. (1995). Mathematics education as a ‚design science'. *Educational Studies in Mathematics, 29*, 355–374. (Cited on p. 6.)

Xie, J., & Masingila, J. O. (2017). Examining interactions between problem posing and problem solving with prospective primary teachers: a case of using fractions. *Educational Studies in Mathematics, 96*(1), 101–118. (Cited on pp. 5, 57, 75, 80, 115, 126, 127, 148, 154, 163.)

Yaglom, I. M. (1962). *Geometric Transformations I.* The Mathematical Association of America. (Cited on p. 26.)

Yeo, J. B. W. (2017). Development of a framework to characterise the openness of mathematical tasks. *International Journal of Science and Mathematics Education, 15*(1), 175–191. (Cited on pp. 96, 97, 100, 104, 105, 108, 109, 229.)

Yimer, A., & Ellerton, N. F. (2010). A five-phase model for mathematical problem solving: Identifying synergies in pre-service-teachers' metacognitive and cognitive actions. *ZDM – Mathematics Education, 42*(2), 245–261. (Cited on pp. 165, 168, 191, 197.)

Yuan, X., & Sriraman, B. (2011). An exploratory study of relationships between students' creativity and mathematical problem-posing abilities. In B. Sriraman & K. H. Lee (Eds.), *The Elements of Creativity and Giftedness in Mathematics* (pp. 5–28). Sense Publishers. (Cited on pp. 4, 82, 160, 164, 192.)

Ziegenbalg, J. (2014). *Elementare Zahlentheorie. Beispiele, Geschichte, Algorithmen* (2nd ed.). Springer. (Cited on p. 57.)

Zimmerman, B. J., & Moylan, A. R. (2009). Self-Regulation. Where metacognition and motivation intersect. In J. Hacker Douglas, J. Dunlosky, & A. C. Graesser (Eds.), *Handbook of Metacognition in Education* (pp. 299–315). Routledge. (Cited on pp. 195, 222.)

List of Tables

L. Baumanns, *Mathematical Problem Posing*, Kölner Beiträge zur Didaktik der Mathematik, https://doi.org/10.1007/978-3-658-39917-7

List of Figures

L. Baumanns, *Mathematical Problem Posing*, Kölner Beiträge zur Didaktik der Mathematik, https://doi.org/10.1007/978-3-658-39917-7

Part VI

Appendix

Online supplement of journal article 2

A

A.1 List of all 47 included articles included in the review

Bicer, A., Lee, Y., Perihan, C., Capraro, M. M., & Capraro, R. M. (2020). Considering mathematical creative self-efficacy with problem posing as a measure of mathematical creativity. *Educational Studies in Mathematics, 105*(3), 457–485.

Bonotto, C. (2013). Artifacts as sources for problem-posing activities. *Educational Studies in Mathematics, 83*(1), 37–55.

Cai, J. (1998). An investigation of U.S. and Chinese students' mathematical problem posing and problem solving. *Mathematics Education Research Journal, 10*(1), 37–50.

Cai, J. (2003). Singaporean students' mathematical thinking in problem solving and problem posing: an exploratory study. *International Journal of Mathematical Education in Science and Technology, 34*(5), 719–737.

Cai, J., & Hwang, S. (2002). Generalized and generative thinking in US and Chinese students' mathematical problem solving and problem posing. *The Journal of Mathematical Behavior, 21*(4), 401–421.

L. Baumanns, *Mathematical Problem Posing*, Kölner Beiträge zur Didaktik der Mathematik, https://doi.org/10.1007/978-3-658-39917-7

Cankoy, O. (2014). Interlocked problem posing and children's problem posing performance in free structured situations. *International Journal of Science and Mathematics Education, 12*, 219–238.

Christou, C., Mousoulides, N., Pittalis, M., Pitta-Pantazi, D., & Sriraman, B. (2005). An empirical taxonomy of problem posing processes. *ZDM – Mathematics Education, 37*(3), 149–158.

Crespo, S. (2003). Learning to pose mathematical problems: Exploring changes in preservice teachers' practices. *Educational Studies in Mathematics, 52*(3), 243–270.

Crespo, S., & Sinclair, N. (2008). What makes a problem mathematically interesting? Inviting prospective teachers to pose better problems. *Journal of Mathematics Teacher Education, 11*(5), 395–415.

da Ponte, J. P., & Henriques, A. (2013). Problem posing based on investigation activities by university students. *Educational Studies in Mathematics, 83*(1), 145–156.

Ellerton, N. F. (2013). Engaging pre-service middle-school teacher-education students in mathematical problem posing: development of an active learning framework. *Educational Studies in Mathematics, 83*(1), 87–101.

English, L. D. (1997). The development of fifth-grade children's problem-posing abilities. *Educational Studies in Mathematics, 34*(3), 183–217.

English, L. D. (1998). Children's problem posing within formal and informal contexts. *Journal for Research in Mathematics Education, 29*(1), 83–106.

Gierdien, F. (2009). More than multiplication in a 12x12 multiplication table. *International Journal of Mathematical Education in Science and Technology, 40*(5), 662–669.

Gierdien, F. (2012). Quadratic expressions by means of 'summing all the matchsticks'. *International Journal of Mathematical Education in Science and Technology, 43*(6), 811–818.

Guo, M., Leung, F. K. S., & Hu, X. (2020). Affective determinants of mathematical problem posing: the case of Chinese Miao students. *Educational Studies in Mathematics, 105*(3), 367–387.

Harpen, X. Y. V., & Sriraman, B. (2013). Creativity and mathematical problem posing: an analysis of high school students' mathematical problem posing in China and the USA. *Educational Studies in Mathematics, 82*(2), 201–221.

Haylock, D. (1997). Recognising mathematical creativity in schoolchildren. *ZDM – Mathematics Education, 29*(3), 68–74.

Headrick, L., Wiezel, A., Tarr, G., Zhang, X., Cullicott, C. E., Middleton, J. A., & Jansen, A. (2020). Engagement and affect patterns in high school mathematics classrooms that exhibit spontaneous problem posing: an exploratory framework and study. *Educational Studies in Mathematics, 105*(3), 435–456.

Klein, S., & Leikin, R. (2020). Opening mathematical problems for posing open mathematical tasks: what do teachers do and feel? *Educational Studies in Mathematics, 105*(3), 349–365.

Koichu, B. (2008). If not, what yes? *International Journal of Mathematical Education in Science and Technology, 39*(4), 443–454.

Koichu, B. (2010). On the relationships between (relatively) advanced mathematical knowledge and (relatively) advanced problem-solving behaviours. *International Journal of Mathematical Education in Science and Technology, 41*(2), 257–275.

Koichu, B., & Kontorovich, I. (2013). Dissecting success stories on mathematical problem posing: a case of the Billiard Task. *Educational Studies in Mathematics, 83*(1), 71–86.

Kontorovich, I. (2020). Problem-posing triggers or where do mathematics competition problems come from? *Educational Studies in Mathematics, 105*(3), 389–406.

Kontorovich, I., & Koichu, B. (2016). A case study of an expert problem poser for mathematics competitions. *International Journal of Science and Mathematics Education, 14*, 81–99.

Kontorovich, I., Koichu, B., Leikin, R., & Berman, A. (2012). An exploratory framework for handling the complexity of mathematical problem posing in small groups. *The Journal of Mathematical Behavior, 31*(1), 149–161.

Lavy, I., & Bershadsky, I. (2003). Problem posing via "what if not?" strategy in solid geometry — a case study. *The Journal of Mathematical Behavior, 22*(4), 369–387.

Lavy, I., & Shriki, A. (2010). Engaging in problem posing activities in a dynamic geometry setting and the development of prospective teachers' mathematical knowledge. *The Journal of Mathematical Behavior, 29*(1), 11–24.

Leavy, A., & Hourigan, M. (2019). Posing mathematically worthwhile problems: developing the problem-posing skills of prospective teachers. *Journal of Mathematics Teacher Education.*

Leung, S. S., & Silver, E. A. (1997). The role of task format, mathematics knowledge and creative thinking on the arithmetic problem posing of prospective elementary school teachers. *Mathematics Education Research Journal, 9*(1), 5–24.

Leung, S.-k. S. (2013). Teachers implementing mathematical problem posing in the classroom: challenges and strategies. *Educational Studies in Mathematics, 83*(1), 103–116.

Li, X., Song, N., Hwang, S., & Cai, J. (2020). Learning to teach mathematics through problem posing: teachers' beliefs and performance on problem posing. *Educational Studies in Mathematics, 105*(3), 325–347.

Liu, Q., Liu, J., Cai, J., & Zhang, Z. (2020). The relationship between domain- and task-specific self-efficacy and mathematical problem posing: a large-scale study of eighth-grade students in China. *Educational Studies in Mathematics, 105*(3), 407–431.

Lowrie, T. (2002). Young children posing problems: The influence of teacher invention on the type of problems children pose. *Mathematics Education Research Journal, 14*(2), 87–98.

Martinez-Luaces, V., Fernandez-Plaza, J., Rico, L., & Ruiz-Hildalgo, J. F. (2019). Inverse reformulations of a modelling problem proposed by prospective teachers in Spain. *International Journal of Mathematical Education in Science and Technology, online.*

Schindler, M., & Bakker, A. (2020). Affective field during collaborative problem posing and problem solving: a case study. *Educational Studies in Mathematics, 105*(3), 303–324.

Silber, S., & Cai, J. (2017). Pre-service teacher's free and structured mathematical problem posing. *International Journal of Mathematical Education in Science and Technology, 48*(2), 163–184.

Silver, E., Mamona-Downs, J., Leung, S., & Kenney, P. (1996). Posing mathematical problems: An exploratory study. *Journal for Research in Mathematics Education*, *27*(3), 293–309.

Silver, E. A., & Cai, J. (1996). An analysis of arithmetic problem posing by middle school students. *Journal for Research in Mathematics Education*, *27*(5), 521–539.

Singer, F. M., & Voica, C. (2013). A problem-solving conceptual framework and its implications in designing problem-posing tasks. *Educational Studies in Mathematics*, *83*(1), 9–26.

Singer, F. M., Voica, C., & Pelczer, I. (2017). Cognitive styles in posing geometry problems: implications for assessment of mathematical creativity. *ZDM – Mathematics Education*, *49*(1), 37–52.

Thanheiser, E., Olanoff, D., Hillen, A., Feldman, Z., Tobias, J. M., & Welder, R. M. (2016). Reflective analysis as a tool for task redesign: The case of prospective elementary teachers solving and posing fraction comparison problems. *Journal of Mathematics Teacher Education*, *19*(2), 123–148.

Tichá, M., & Hošpesová, A. (2013). Developing teachers' subject didactic competence through problem posing. *Educational Studies in Mathematics*, *83*(1), 133–143.

Van Harpen, X. Y., & Presmeg, N. C. (2013). An investigation of relationships between students' mathematical problem-posing abilities and their mathematical content knowledge. *Educational Studies in Mathematics*, *83*(1), 117–132.

Voica, C., & Singer, F. M. (2013). Problem modification as a tool for detecting cognitive flexibility in school children. *ZDM – Mathematics Education*, *45*(2), 267–279.

Voica, C., Singer, F. M., & Stan, E. (2020). How are motivation and self-efficacy interacting in problem-solving and problem-posing? *Educational Studies in Mathematics*, *105*(3), 487–517.

Xie, J., & Masingila, J. O. (2017). Examining Interactions between Problem Posing and Problem Solving with Prospective Primary Teachers: A Case of Using Fractions. *Educational Studies in Mathematics*, *96*(1), 101–118.

A.2 Categorisation of all 47 articles within the developed framework

Dimension 1: Generating and reformulating

<table>
<tr><th>Generating</th><th>Reformulating</th></tr>
<tr><td>Bicer et al. (2020)
Bonotto (2013)
Cankoy (2014)
Christou et al. (2005)
Crespo (2003)
Crespo and Sinclair (2008)
English (1998)
Gierdien (2009)
Gierdien (2012)
Guo et al. (2020)
Kontorovich et al. (2012)
Leavy and Hourigan (2019)
Lowrie (2002)
Liu et al. (2020)
E. A. Silver and Cai (1996)
E. Silver et al. (1996)
Singer and Voica (2013)
Tichá and Hošpesová (2013)
Voica et al. (2020)</td><td>Ellerton (2013)
Klein and Leikin (2020)
Koichu (2008)
Koichu (2010)
Lavy and Shriki (2010)
Martinez-Luaces et al. (2019)
da Ponte and Henriques (2013)
Thanheiser et al. (2016)</td></tr>
<tr><td colspan="2">English (1997)
Haylock (1997)
Koichu and Kontorovich (2013)
Kontorovich and Koichu (2016)
Kontorovich (2020)
Lavy and Bershadsky (2003)
S. S. Leung and Silver (1997)
Li et al. (2020)
Silber and Cai (2017)
Singer et al. (2017)
Van Harpen and Presmeg (2013)
Harpen and Sriraman (2013)
Voica and Singer (2013)
Xie and Masingila (2017)</td></tr>
</table>

Dimension 2: Routine and non-routine problems

<table>
<tr><th>Routine Problems</th><th>Non-Routine Problems</th></tr>
<tr><td></td><td>Cankoy (2014)</td></tr>
<tr><td></td><td>Cai (1998)</td></tr>
<tr><td></td><td>Ellerton (2013)</td></tr>
<tr><td></td><td>English (1998)</td></tr>
<tr><td></td><td>Gierdien (2009)</td></tr>
<tr><td></td><td>Gierdien (2012)</td></tr>
<tr><td></td><td>Guo et al. (2020)</td></tr>
<tr><td></td><td>Klein and Leikin (2020)</td></tr>
<tr><td></td><td>Koichu (2008)</td></tr>
<tr><td>Bicer et al. (2020)</td><td>Koichu (2010)</td></tr>
<tr><td>Bonotto (2013)</td><td>Koichu and Kontorovich (2013)</td></tr>
<tr><td>Christou et al. (2005)</td><td>Kontorovich et al. (2012)</td></tr>
<tr><td>Crespo and Sinclair (2008)</td><td>Kontorovich (2020)</td></tr>
<tr><td>S.-k. S. Leung (2013)</td><td>Kontorovich and Koichu (2016)</td></tr>
<tr><td>Thanheiser et al. (2016)</td><td>Lavy and Bershadsky (2003)</td></tr>
<tr><td></td><td>Lavy and Shriki (2010)</td></tr>
<tr><td></td><td>Leavy and Hourigan (2019)</td></tr>
<tr><td></td><td>Li et al. (2020)</td></tr>
<tr><td></td><td>Martinez-Luaces et al. (2019)</td></tr>
<tr><td></td><td>Schindler and Bakker (2020)</td></tr>
<tr><td></td><td>Silber and Cai (2017)</td></tr>
<tr><td></td><td>E. Silver et al. (1996)</td></tr>
<tr><td></td><td>Singer et al. (2017)</td></tr>
<tr><td></td><td>Voica et al. (2020)</td></tr>
<tr><td colspan="2">Cai (2003)</td></tr>
<tr><td colspan="2">Cai and Hwang (2002)</td></tr>
<tr><td colspan="2">Crespo (2003)</td></tr>
<tr><td colspan="2">English (1997)</td></tr>
<tr><td colspan="2">Haylock (1997)</td></tr>
<tr><td colspan="2">Headrick et al. (2020)</td></tr>
<tr><td colspan="2">S. S. Leung and Silver (1997)</td></tr>
<tr><td colspan="2">Lowrie (2002)</td></tr>
<tr><td colspan="2">Liu et al. (2020)</td></tr>
<tr><td colspan="2">da Ponte and Henriques (2013)</td></tr>
<tr><td colspan="2">Singer and Voica (2013)</td></tr>
<tr><td colspan="2">Tichá and Hošpesová (2013)</td></tr>
<tr><td colspan="2">Van Harpen and Presmeg (2013)</td></tr>
<tr><td colspan="2">Harpen and Sriraman (2013)</td></tr>
<tr><td colspan="2">Voica and Singer (2013)</td></tr>
<tr><td colspan="2">Xie and Masingila (2017)</td></tr>
</table>

Dimension 3: Metacognitive behaviour

Low metacognitive behaviour	High metacognitive behaviour
	Bonotto (2013) Cankoy (2014) Cai and Hwang (2002) Cai (2003) Crespo and Sinclair (2008) Ellerton (2013) English (1997) Gierdien (2009) Gierdien (2012) Klein and Leikin (2020) Koichu (2008) Koichu (2010) Koichu and Kontorovich (2013) Kontorovich et al. (2012) Kontorovich and Koichu (2016) Kontorovich (2016) Lavy and Shriki (2010) Leavy and Hourigan (2019) Lowrie (2002) Liu et al. (2020) Martinez-Luaces et al. (2019) da Ponte and Henriques (2013) Silber and Cai (2017) E. Silver et al. (1996) Singer and Voica (2013) Thanheiser et al. (2016) Voica and Singer (2013) Voica et al. (2020) Xie and Masingila (2017)
Cai (1998) Crespo (2003) Lavy and Bershadsky (2003) Tichá and Hošpesová (2013)	

B

Online supplement of journal article 3

Anchor examples for the developed episode types

In addition to the examples of episode types listed in the article, in this supplement, we provide further anchor examples. This should make it easier for readers to understand the episode types and apply them themselves. The anchor examples come from different processes on the situation Nim game. In order to better situate the anchor examples in their processes and to be able to form an unbiased picture of them, we provide a complete interview narrative of the processes in which the anchor examples are located before showing the coded anchor examples. The anchor examples described below are taken from three different processes. After each paragraph in these descriptions, we indicate the amount of time spent on the paragraph described.

Interview narratives of the processes used to extract the anchor examples

Xanthippe & Yasmin

Xanthippe and Yasmin use working backwards to find a winning strategy for player A after a short time, i.e., 1:30 minutes. They realize that player B has no chance of winning the Nim game when there are 6 stones on the table. One by one, they increase the number of stones on the table.

L. Baumanns, *Mathematical Problem Posing*, Kölner Beiträge zur Didaktik der Mathematik, https://doi.org/10.1007/978-3-658-39917-7

They realize that even with 12 stones, player B cannot win. Finally, they work out the solution on their own. (Duration: 00:00–22:45)

After solving the problem, at the beginning of their problem-posing process, they first increase the number of stones that are on the table. Numbers 21 to 30 come up, and they agree on 25 and ask the question: "How do the chances of winning behave with 25 stones?" They also ask the same question for 30 and 50 stones. In doing so, they reflect that it is good to also have a number with 30 that is part of the row of three. (Duration: 22:45–23:56)

Then they change the number of stones that can be removed from the table. Yasmin suggests that you can remove 1, 2, or 3 stones. Xanthippe states that removing 2 or 3 stones could also be interesting. (Duration: 23:56–25:29)

Both then first think about a solution for the variation where you may remove 1, 2, or 3 stones. They realize that the winning strategy here remains similar, but the row of four is decisive. This idea is further generalized for removing 1 to 4 stones. Subsequently, the previous ideas are written down. In the process, they talk a lot about the exact formulation of their tasks. (Duration: 25:29–27:40)

Xanthippe then introduces the idea of how the chance of winning changes in a Nim game with 3, 4, or 5 players. The two do not pursue the solution of this task any further. Yasmin suggests that you could also ask how this game would behave with magnets without elaborating on this idea. (Duration: 27:40–28:49)

Yasmin then reflects on what variables the Nim game contains, which ones have already been changed, and which ones could be changed for more variations. Xanthippe explains that she doesn't know what else could be changed and suggests that a story could be developed around this game to help students solve the initial problem. These thoughts are not elaborated on. Yasmin explains that they are finished. (Duration: 28:49–31:10)

Tino & Ulrich

Tino and Ulrich first work out a winning strategy for the Nim game. Their heurism is to work backwards right at the beginning. They consider that with 3 stones, player B has no more chance to win and they then extend this idea to 6 stones. They thereafter generalize this idea for the 20 stones and formulate the solution afterward. After that, they move on to problem posing on their own. (Duration: 00:00–07:45)

Ulrich suggests right at the beginning to pose simple variations of the game, for example, if you increase the number of stones to 21. Ulrich says, however, that it would be more exciting if they varied the number of stones they are allowed to remove instead. He suggests the rule of removing 1, 2, or 3 stones. Tino writes down the ideas and both also work on the formulation of the respective tasks. (Duration: 07:45–09:31)

Then they play through the variation of removing 1, 2, or 3 stones. When playing, they do not yet achieve sufficient certainty for a solution strategy. They then work backwards; however, they still do not work out a final solution, but then want to move on to other ideas for problems. (Duration: 09:31–15:52)

They now expand the rules for winning the game by adding that you must take the last stone and also have more stones removed from the table after the last move. To work out a winning strategy for this variation of the game, they play it again with this new rule. However, after some time they turn back to the solution for removing 1, 2, or 3 stones. They reach a solution, and after their initial failure, evaluate it positively for a math game if there is a safe winning strategy. (Duration: 15:52–24:32)

After Ulrich briefly tries to further generalize a possible winning strategy for the case where you are allowed to remove 1 to 4 stones, he suggests another idea: you are only allowed to remove 2 or 3 stones from the table. However, Tino first needs a longer time to write down the formulations for the games posed so far. (Duration: 24:32–26:48)

Afterward, they notice that in the game where you are only allowed to remove 2 or 3 stones, you can get to the point where no move is possible because there is only 1 stone left on the table. They consider whether such an outcome also leads to losing the game or whether this outcome means a draw. Without coming to a mutual answer on this question, they consider whether there is a safe winning strategy for this variation. They are converging on a solution to this question, but do not come to a completely satisfactory conclusion. They then consider what changes if you are only allowed to remove 3 or 4 stones from the table. Afterward, Tino again takes the opportunity to write down all the ideas collected so far. (Duration: 26:48–34:18)

Ulrich expands the rules of the game and suggests that you get another victory point if the number of stones you have removed from the table in a round is divisible by three. They think through this idea by playing a round of this game. Ulrich then moves further away from the Nim game with his ideas, suggesting that each player starts with 10 stones with which to build a square pyramid. Without posing a final problem to this idea, the process ends. (Duration: 34:18–39:39)

Noam & Oskar

Noam and Oskar, when solving the problem, initially believe that it could play a role in a winning strategy whether the number of stones is even or odd. After some thought and help from the interviewer to work backwards, Noam and Oskar found the correct solution. (Duration: 00:00–19:00)

At the beginning of the subsequent problem-posing process, Noam and Oskar first pose the more general problem of how many stones must be on the table for player A to win safely. They express liking the problem because it could help figuring out the winning strategy. Then they pose the variation of the Nim game where you are allowed to remove 1, 2, or 3 stones from the table. After initial difficulties on solving this variation, Noam expresses the idea that here the players have to get to a number divisible by four to win for sure. They then generalize this idea to the case where you are allowed to remove 1 to n stones from the table. They test out their developed winning strategies by playing the Nim game together with the corresponding modifications. (Duration: 19:00–24:52)

Then they increase the number of stones that are on the table. Specifically, the number 100 is called, and later prime numbers such as 47. Subsequently, the previous ideas are formulated as problems. Oskar then asks Noam again why they thought parity would play a role in the winning strategy when solving the task at the beginning. (Duration: 24:52–25:55)

Oskar stacks the stones to a tower (2 – 1 – 2 – 1 – ...) and adds as an additional rule, you may only remove stones from one row. Noam brings in an idea for an auxiliary task: You could simplify the game and ask who wins if you are only allowed to remove one stone at a time. Then, they both realize, divisibility by two, i.e. parity, plays the central role in the winning strategy. (Duration: 25:55–27:33)

Then both change the rules to allow player A to remove a maximum of 3 stones and player B to remove a maximum of 2 stones from the table. While playing this game, they realize that player A, as the starting player with a winning strategy, may now enjoy only one more advantage.

Therefore, they change this rule and ask themselves what happens if player B is allowed to remove a maximum of 3 and player A a maximum of 2 stones from the table. (Duration: 27:33–29:14)

They play through several rounds of this game. In the process, they express in between whether they like the game. However, in the end, they are not satisfied with this idea. (Duration: 29:14–32:33)

Oskar finally picks up the idea of stacking the stones that he had posed earlier. This time he suggests stacking the stones in a square pyramid. He again adopts the idea that you can only remove stones from one level of the pyramid. They both note that this rule makes it necessary to consider whether or not to start as a player at each level of the pyramid. This new twist on the game appeals to both of them. Their trial ends on this note. (Duration: 32:33–36:06)

Anchor examples

Situation Analysis

In the following anchor example, Xanthippe (X) and Yasmin (Y) are at the end of a longer phase of variation within their problem-posing process initiated by the Nim game. The following example arises in a moment of lack of ideas, in which they consider which conditions of the initial problem they have not yet varied.

Anchor example (Duration: 28:49–29:29)

X: Are there any other options? Well, variables that could really be affected would really be the number of players, the number of stones, how many you take away.

Y: The number of steps, exactly. The only thing you could do now is to put it into another context, into another everyday context. So, the only thing I can think of is to somehow make a story around it.

Variation

In the following example, Tino (T) and Ulrich (U) are at the very beginning of their process. Tino instantaneously starts varying the number of stones on the table in the Nim game. Then he varies the condition that you may remove one or two stones from the table.

Anchor example (Duration: 07:54–08:35)

T: The first one we are going to ask now [...] is, we change the number of stones. That is like an easy variation task in the textbook. Instead of 20 we now have 21 stones. Can player A win? No, he can't.

U: Exactly, then basically player B can win for sure. That essentially turns the whole thing around [...].

T: And the next thing before we write it down, I would like to play it: You can take 1, 2 or 3 stones. That makes the whole thing much more exciting.

Generation

After Noam (N) and Oskar (O) run out of ideas for further variations in the following anchor example, Noam constructs more conditions to add to the Nim game. He stacks the given stones pyramidically. In the lowest level there are 4x4 stones, above them 3x3, then 2x2 and finally one stone on top. However, on your turn, you may only remove stones from one level.

Anchor example (Duration: 33:05–35:02)

O: (stacks the stones to a pyramid) Or I think it's a really cool idea if you have some kind of shape, like [...] Mahjong [...] where you [...] just take the stones that are sort of exposed [...]. I think that's not such a stupid game. You can take up to 3 stones, but not stones from different levels, so you can only take stones from one level. That means the first move is fixed with one and the next move you can decide whether to take one or two.

N: And who wins?

O: Whoever makes the last move.

Problem Solving

In the following anchor example, Tino and Ulrich solve a problem that they posed in the example shown above for variation. The players of the Nim game are allowed to remove 1, 2 or 3 stones. They mistakenly think that they already have this variation under control for situations where 4 or 7 pieces are on the table.

Anchor example (Duration: 21:23–22:30)

U: If the other player has 7 ... oh no, then he could take the 3.

T: Yes, that's why I was confused why you came up with 7, because before it was always a row of three [for standard Nim game], wasn't it? And now our guess was that this is the row of four. Now we just have to test if it really works. So, if you still have 4 at the end, you have won analogous to 3 [for standard Nim game], that means one move before [...] if I still have 8 here, I have to lose [...].

U: And that's the case, because the other one . . . no matter what you do, here 8 is what I believe.

T: Yes, it's simply always the row of four, the maximum number of stones you can move plus 1.

U: That would be a good thesis. And then player A would have lost in any case, because there are 20 stones.

Evaluation

The expressions of the following anchor example directly follow the generated idea developed above in the anchor example for generation in Noam and Oskar's process. There, evaluative statements can already be found in partial sentences. In the following example, Oskar is more specific in this respect.

Anchor example (Duration: 35:12–35:34)

O: But now you just have to think, ok would you like ... at the bottom layer there are 16 stones, would you like to start there or not? Then there are 9 stones above it, ok. Then you just have to decide ok, if there are 9 stones, do you want to start there or not. But I think that's actually cool. Because then you still have to think about the number of stones. That's actually a cool game, because you have to apply the rules differently.

N: Yes. It's really good!

Example of calculating interrater agreement

First, the process of Xerxes and Yuna is described. Like the process in section 4 of the article, this process is already divided into episodes without labelling them with the episode types. The description as well as the division comes from the authors (Rater 1). In a later section, the differences in coding to Rater 2 will be presented. After that, the calculation of the interrater agreement using EasyDIAg is explained (Holle & Rein, 2013).

Process description by Rater 1

Xerxes and Yuna have not been able to work out a correct solution to the Nim game. They justify by working backwards that the game is decided when there are three stones left on the table. They also formulate the justified hypothesis that one player can react to the moves of the other player. They believe that the probability of winning depends on the parity of the number of stones that are on the table at the beginning. However, the solution finally is given by the interviewer.

Episode 1 (00:00–00:33): At the beginning, Yuna varies the rule that you remove stones and wants to create something where you add stones. However, this idea is not further developed.

Episode 2 (00:33–01:10): Then Xerxes and Yuna analyze again the winning strategy presented by the interviewer.

Episode 3 (01:10–02:27): As a new idea, Yuna suggests that players must make sure that they have an odd number of stones at the end of the game. Xerxes does not understand this idea at first. Yuna repeats her idea.

Episode 4 (02:27–04:04): Both try to understand the game with the newly added rule by playing through it several times. For clarity, they decide to play with eleven pieces. At first, Yuna expresses that the game is not meaningful because player A always wins.

Episode 5 (04:04–04:49): Yuna suggests that one stone be added back to the table after each turn. Xerxes notes that this would become an endless game. Yuna then suggests the idea of removing only 2 or 3 stones or 1 or 3 stones from the table.

Episode 6 (04:49–05:28): Then Yuna writes down these and previous ideas.

Episode 7 (05:28–07:04): The game, in which you may only remove 1 or 3 pieces from the table, is then played in a round with 20 starting stones. They don't actually finish the game. Instead, they consider possible moves that could lead to victory. They subsequently express uncertainty as to whether their game is predictable, so that one reaches a sure winning strategy.

Episode 8 (07:04–08:00): They then pick up a previously developed idea again. Their new game is that you can take away 1 or 2 stones, but after each turn one stone is added.

Episode 9 (08:00–08:40): Again, they first make a general statement that you believe that one player must always win. They concretize their thoughts by making exemplary plays. In doing so, they realize, as previously suspected, that it could then be an endless game.

Episode 10 (08:40–09:33): Yuna then continues to vary the condition of the Nim game as to how many stones may be removed from the table. She modifies this rule by using the Fibonacci sequence. One may first remove only one stone from the table, then up to two, then up to three, up to five, and so on. She further suggests that player A may always remove only an odd number of stones while player B may always remove only an even number of stones.

Episode 11 (09:33–10:28): They both try to understand the last-mentioned idea in more depth by discussing a possible scenario of this game. However, Xerxes and Yuna do not work out a concrete winning strategy.

Episode 12 (10:28–11:05): Yuna and Xerxes digress from the task at hand. Yuna jokingly suggests that the task could be to throw the stones into a basket.

Episode 13 (11:05–13:00): Yuna notes that all variations have now been worked through. Yuna then arranges the stones in a 3x3 field. They try to match the character of Tic-Tac-Toe and Connect Four with that of the Nim game. However, they do not arrive at a new game that is satisfactory to them, so they subsequently end the process.

Process coding by Rater 1

Yuna's initial attempts in episode 1 to vary a rule of the game were coded as variation. The subsequent exploration of the winning strategy presented by the interviewer was coded as analysis. In episode 3, Yuna adds a new condition. Accordingly, this episode was coded as a generation. The fact that Xerxes asks what exactly Yuna means by her added condition reinforces the interpretation that new and different conditions from the initial problem were added. In episode 4, they both pursue problem solving. Yuna's assessment that no meaningful problem has emerged is followed by the joint coding of problem solving & evaluation. Since in episode 4 a specific condition of the initial problem is again modified by Yuna, this episode was coded as a variation. Episode 6 was coded as non-content related episode writing. Although they do not come to a satisfactory conclusion, Xerxes and Yuna engage in problem solving in Episode 7. In doing so, they evaluate that their problem is probably not well-defined in the sense that one can work out a winning strategy. Therefore, this episode was coded as problem solving & evaluation. In episode 8, they pick up an idea from episode 1, where stones are added back to the table. Analogous to episode 1, this episode is also coded as a variation. Once again, they try to solve the problem judge it as an endless game. This episode 9 was coded as problem solving & evaluation. The two new games posed in episode 10 by Yuna emerge from a variation as specific conditions of the Nim game are modified. In Episode 11, Xerxes and Yuna focus on problem solving of the second game they posed in the previous episode. In episode 12, it is noticeable that their concentration is already waning because they are no longer generating ideas. Their humorous suggestions were coded as non-content related episode digression. Yuna's goal in the last episode is to move further away from the Nim game by incorporating ideas and rules from other games. This rather superficial attempt was coded as generation.

Differences between the codings by rater 1 and rater 2

The differences between the coding of Rater 1 in the previous section and that of Rater 2 are described below. The focus is on those differences where the agreement table calculated by EasyDIAg produced entries outside the main diagonal at an overlap parameter of 60% (see Table B.1 on page 301). Namely, these entries reduce the interrater agreement of the coding manual used, as calculated in the present study. A graphical comparison of both codings can be found in Figure B.1 on the following page.

In the present process of Xerxes and Yuna, there are two time periods where the codings of Rater 1 and Rater 2 differ. The first difference concerns episode 8 of the coding of Rater 1. Rater 1 has coded a variation here. Rater 2 coded problem solving & evaluation throughout, as in the

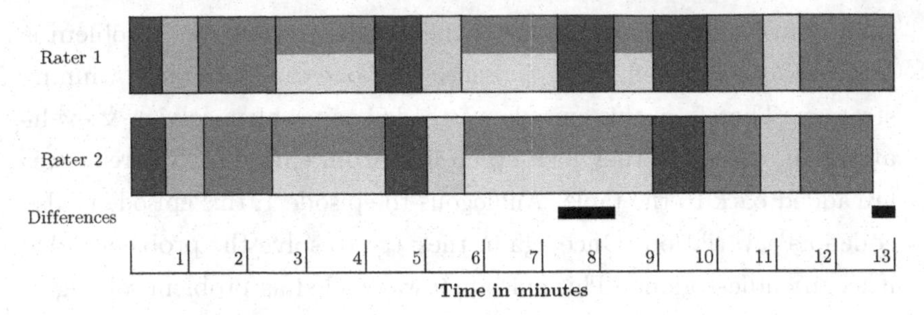

Fig. B.1.: Comparison of the time-line charts of the problem-posing process as coded by Rater 1 and Rater 2.

previous and subsequent episodes. Rater 2 justifies this with the fact that the game Yuna posed in this time period was already formulated in episode 1. The superordinate goal of this time period is still to solve the problems that are posed and to evaluate them.

A second difference between the codings of Rater 1 and Rater 2 concerns the end of the process. While Rater 1 codes the end of the process at 13:00 min, Rater 2 sets it already at 12:25 min. In fact, in the remaining 35 seconds that Rater 1 still coded as generation, there are numerous behaviors, especially in Xerxes, that do not reveal any actual content-related engagement with the Nim game. Yuna, on the other hand, still seems to be mentally engaged with new games.

Calculating values for agreement table

From the presented coding of Rater 1 and the differences in the coding of Rater 2, results the agreement table in Table B.1 on the next page calculated by EasyDIAg. On the main diagonal are the matching codings. The value of the number in the individual cells of the table results from the number of set codes as well as their temporal expansion. In the following, we will refer to entries of the table in which, for example, Rater 1 coded Evaluation and Rater 2 coded Problem Solving as (E |

PS). The differences described before can be found in the table in the entries (V | PS/E) and (G | X). (V | PS/E) refers to the 56 seconds time segment in which Rater 1 coded variation while Rater 2 coded problem solving & evaluation. (G | X) refers to the last 25 seconds in which Rater 1 coded generation while Rater 2 already coded the end of the process. This table now results in a Cohen's Kappa of $\kappa = .84$.

Rater 2 \ Rater 1	SA	V	G	PS	PS/E	E	O	X	Total	p
SA	2	0	0	0	0	0	0	0	2	0.08
V	0	6	0	0	0	0	0	0	6	0.24
G	0	2	4	0	0	0	0	0	4	0.16
PS	0	0	0	2	0	0	0	0	2	0.08
PS/E	0	0	2	2	2	0	0	0	4	0.16
E	0	0	0	0	0	0	0	10	0	0.00
O	0	0	0	0	0	0	4	0	4	0.17
X	0	0	1	0	0	0	0	–	1	0.04
Total	2	8	5	2	2	0	4	0	23	1.0
p	0.08	0.32	0.20	0.08	0.08	0.00	0.17	0.00	1.00	**0.84**

Tab. B.1.: Agreement table for all seven categories of episodes as determined by EasyDIAg. The $\%_{overlap}$ parameter was set to 60%. (Abbreviations: SA = Situation Analysis, V = Variation, G = Generation, PS = Problem Solving, PS/E = Problem Solving & Evaluation [simultaneous coding], E = Evaluation, O = Others, X = No Match)

Printed in the United States
by Baker & Taylor Publisher Services